MATRIX ISOLATION

MATRIX ISOLATION

A technique for the study of reactive inorganic species

STEPHEN CRADOCK
Lecturer in Chemistry, University of Edinburgh

A. J. HINCHCLIFFE
Officer in the Instructor Branch, Royal Navy

CAMBRIDGE UNIVERSITY PRESS
Cambridge
London · New York · Melbourne

Published by the Syndics of the Cambridge University Press
The Pitt Building, Trumpington Street, Cambridge CB2 1RP
Bentley House, 200 Euston Road, London NW1 2DB
32 East 57th Street, New York, NY 10022, USA
296 Beaconsfield Parade, Middle Park, Melbourne 3206, Australia

First published 1975

Printed in Great Britain at the
University Printing House, Cambridge
(Euan Phillips, University Printer)

Library of Congress cataloguing in publication data

Cradock, Stephen
Matrix isolation

Bibliography: p. 141
Includes index
1. Matrix isolation spectroscopy
I. Hinchcliffe, A. J., joint author. II. Title
QD96.M33C7 541'.223 74-31786
ISBN 0 521 20759 2

Contents

Preface

Matrix isolation is a technique developed over the last two decades which enables reactive, short-lived molecules to be trapped in a solid matrix and studied spectroscopically. This book aims to show how this may be done in practice. It assumes that the reader is familiar with much basic physics, chemistry and spectroscopy, but the nature and formation of the matrix and details of its interactions with trapped species are covered explicitly. There are also chapters on various technical matters relevant to the production and study of matrix-isolated samples.

A major portion of the book consists of specific examples where study of matrix-isolated samples has contributed to our understanding of the nature of small, reactive molecules. As well as 'free radicals' and the like we have included what may be termed 'high temperature monomers', species such as sodium chloride that are normally met only in a polymeric form. The monomeric molecules can be studied in matrices and information concerning bonding and structure obtained. We have attempted to maintain a critical stance in discussing the examples, so as to bring out clearly the physical and chemical significance of the results, as well as to indicate where the interpretation of the results is less than certain.

In the examples we have interpreted the term 'inorganic' rather widely, including a number of small carbon-centred species, as it seemed pointless to exclude derivatives of one element arbitrarily. On the other hand we have barely touched on the vast field of study of organic (and more recently 'hetero-organic') radicals in organic glassy matrices.

We would like to thank Professor E. A. V. Ebsworth for his support and encouragement during the preparation of the manuscript, and for several constructive criticisms of our presentation of examples.

S.C.
A.J.H.

January 1975

1 Introduction

The chemist is usually concerned to know the structure and other properties of individual molecules, yet matter is rarely found in the form of isolated molecules. Intermolecular interactions dominate the physical nature of matter in the solid and liquid phases, and are experimentally observable even in gases, where they are smallest. In general the 'molecular' properties of a substance can only be deduced from gas-phase studies. The intermolecular interactions are strongest between chemically reactive species such as most atoms, free radicals and 'high temperature monomers', all of which can be studied in the gas phase only at low concentrations and high effective temperatures. Even under such extreme conditions some species are so reactive that they exist for only a few micro- or milliseconds after they are formed, so that the study of their molecular properties is a difficult matter.

The technique of matrix isolation is one result of attempts to overcome some of the difficulties associated with the study of very reactive molecules. In essence, the method involves the trapping of the molecule in a *rigid cage* of a *chemically inert* substance (the *matrix*) at a *low temperature*. The rigidity of the cage prevents diffusion of reactive molecules, which would lead to reaction with other such species. The inertness of the matrix material prevents loss of reactive molecules by reaction with their environment. The low temperature, besides contributing to the rigidity of the cage, serves to reduce the rate of possible internal rearrangements that require any activation energy. Under such conditions molecules that normally have very short lifetimes can be preserved indefinitely and studied at leisure.

In practice, few materials other than the rare gases and molecular nitrogen are chemically inert enough to serve as matrices for the most reactive species. The formation of a rigid matrix implies the use of temperatures not exceeding about one-third of the melting point of the solid, i.e. temperatures of 9 K for neon, 29 K for argon, 40 K for krypton, 55 K for xenon or 26 K for nitrogen. As the lowest temperature attainable using liquid nitrogen as coolant is 63 K, the triple point of nitrogen, the most inert materials available can only be used as matrices if colder refrigerants are employed. Only liquid hydrogen and liquid helium are suitable; they are usable over the ranges 12–33 K and 2–5 K under 'boil off' pressures that can be controlled to adjust the temperature of the liquid. The necessity for the use of such low temperatures has controlled the development of the technique of matrix isolation.

The first experiments recognisably related to our subject were in fact carried out in 1924 in the pioneering cryogenic (low temperature) research laboratory of Kamerlingh Onnes in Leiden. Vegard studied the emission spectra of oxygen and nitrogen atoms produced by electron, proton or X-ray bombardment of impure solid nitrogen or solid mixtures of nitrogen and rare gases, using liquid hydrogen and liquid helium as refrigerants. These were not available elsewhere at that time, and the experiments were not repeated and extended until about thirty years later. In the early 1950s Broida in Washington and Pimentel in Berkeley began to use the matrix isolation technique in the study of atoms and reactive molecules, but the method spread only slowly until the wider availability of liquid helium in the early 1960s and the advent of microrefrigerators in the last few years made it possible for matrix isolation experiments to be performed outside the United States or the immediate neighbourhood of physics departments with surplus liquid helium. The experimental procedure has become so convenient and cheap that argon or nitrogen matrices at 4 K are increasingly used in the study of stable molecules although the low temperature and complete chemical inertness are strictly unnecessary.

The matrix isolation technique necessarily involves a combination of several distinct technologies, each of which interacts with the others. The most basic factor, the low temperature needed to give rigid matrices, implies cryogenic technology, and in turn requires the use of high vacuum techniques without which low temperatures cannot conveniently be maintained. The nature of the matrix, the low temperature and the need to isolate the sample in a vacuum all imply that only spectroscopic methods can be used to study matrix-isolated species, and the experimental technique is to a large extent dominated by the need to expose the sample to the spectrometer at the same time as cooling it in a high vacuum.

The main spectroscopic methods used to study matrix-isolated species are electronic absorption and emission spectroscopy in the visible and ultraviolet regions, vibrational absorption spectroscopy in the infrared and electron spin resonance (e.s.r.) spectroscopy. Electronic and vibrational absorption studies are usually carried out on samples deposited on cooled windows transparent to the radiation concerned (fig. 1.1(*a*)). Electronic emission spectroscopy, which involves excitation by an intense source of radiation while the spectrometer 'views' the excited sample, is better carried out with the sample deposited on a metal surface (fig. 1.1(*b*)). This type of sample arrangement is also suitable for Raman spectroscopy.

E.s.r. spectroscopy, where the sample is contained in a 'resonant cavity' in a strong magnetic field and irradiated with radiofrequency waves, is often carried out with the sample deposited on a synthetic sapphire rod or plate (fig. 1.1(*c*)). In each case heat transfer from the window, metal surface or rod to the coolant reservoir must be as efficient as possible. This is usually ensured by the use of copper or brass connecting parts which must be as short as possible. The sample sample holder and coolant reservoir must be contained in a vacuum vessel, so windows are provided for ingoing and outgoing radiation. There must also be provision for introducing the matrix gas and sample.

The matrix-isolated species is of course not completely free of intermolecular interactions. The magnitude of these is, however, much lower in solid rare gases or nitrogen than in more normal solid or liquid systems, and it is usually possible to regard the effect of the matrix as a perturbation of the molecular property required. The observed spectroscopic properties may then be taken as representative of those of the free molecule.

The arrangement of this book follows the order suggested in the paragraphs above. The properties of matrix materials are dealt with in chapter 2, chapter 3 covers some relevant aspects of low temperature, high vacuum and oven technology, while the methods by which matrix-isolated reactive species may be prepared are described in chapter 4. The spectroscopic methods used in the study of matrix-isolated substances and the effects of the matrix on the observed spectroscopic properties are discussed in chapters 5 and 6 respectively. Two chapters of examples of systems studied by matrix isolation serve to illustrate the wide scope of the method, and we end with a brief attempt to sum up the achievements of matrix isolation to date and to indicate its further potential.

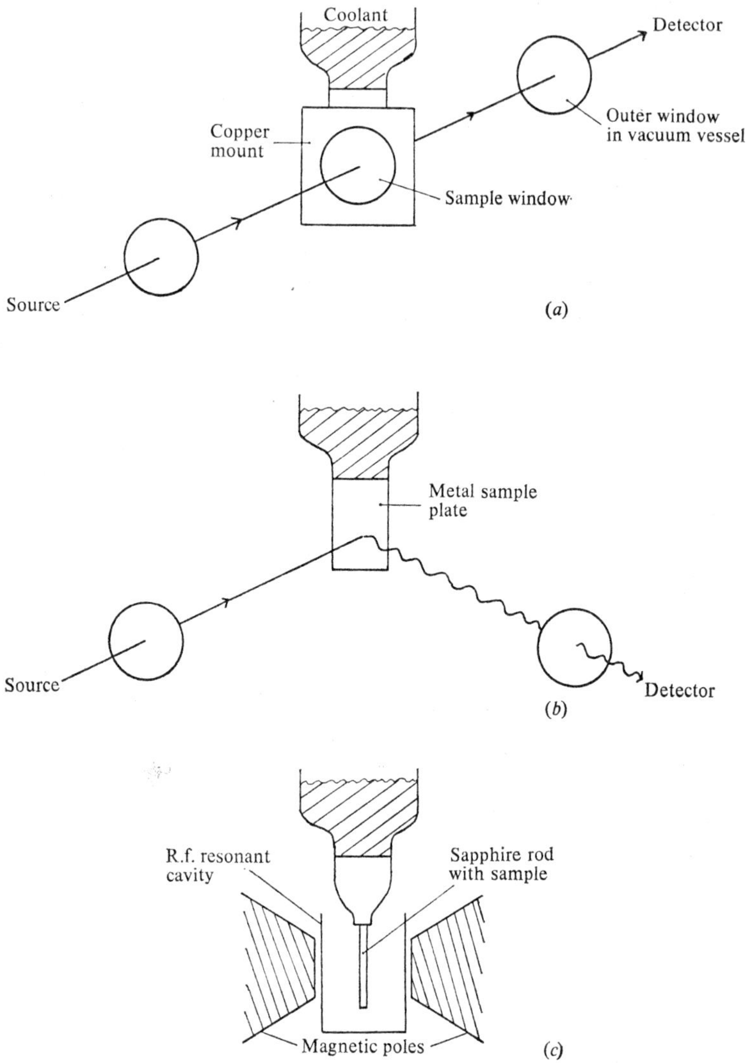

Fig. 1.1. Typical experimental arrangements for spectroscopic studies of matrix-isolated samples. (*a*) Absorption experiment, (*b*) emission experiment, (*c*) e.s.r. experiment.

2 Matrix materials and their properties

As we noted in chapter 1, most matrix isolation of reactive species is done using solid rare gases or nitrogen as matrix material because of the high degree of chemical inertness of these materials. They are also almost uniquely free of absorption spectra that would interfere with spectroscopic detection of the isolated species. Argon and nitrogen are readily and cheaply available, being obtained in large quantities by the fractional distillation of liquid air. The other rare gases are more expensive, being present in the atmosphere only in very small amounts. Only helium is found in relatively concentrated form, in some natural gas samples, from which it is extracted on a large scale in the United States.

There are, however, other potential matrix forming materials, and before going on to discuss the structures and properties of matrices in general we shall briefly list some of the disadvantages of these alternatives.

2.1 Other potential matrix materials

Helium and hydrogen. These may be considered together because, despite their obvious suitability on the grounds of lack of interfering spectra and chemical inertness in all or most likely circumstances, they are quite unusable because they do not form rigid solids and have high vapour pressures even at 4 K. Helium does not solidify at all at pressures less than 25 atmospheres, and hydrogen melts near 14 K. The very low boiling points and high vapour pressures make it impossible to maintain high vacuum conditions in the presence of the condensed phases. This in turn makes it very difficult to maintain low temperatures without elaborate insulation, because, as we shall discuss in chapter 3, heat is transferred across a gas-filled space much faster than across an evacuated space. We include some of the properties of helium and hydrogen in table 2.2 for comparison.

Oxygen, fluorine, chlorine. Although these substances have no interfering absorption bands in the infrared spectrum they are not much used because of their high chemical reactivity. A few experiments using oxygen as matrix material have been reported. In most of these the chemical reactivity was used to advantage, for when an atomic product of photolysis, discharge or evaporation reacts with matrix material to form MO_2 this is more readily detected and identified than the original atom.

Carbon monoxide. Although this material has a strong infrared absorption it has been much used as a matrix material. Its chemical reactivity is not much greater than that of nitrogen, with which it is isoelectronic. The chemical reactivity is often used to good effect; a large number of new metal carbonyl species have been detected when metal atoms are condensed with a carbon monoxide matrix (see chapter 8). The physical properties of carbon monoxide are similar to those of argon and nitrogen.

Methane and other alkanes. Again, although these substances have infrared absorptions they are sometimes used as matrix materials. Their chemical reactivity is low because of the high bond strength of the C–H and C–C bonds and their low polarity. Branched alkanes of C_5 to C_8 size are particularly widely used in the rather distinct technique of organic matrix isolation, as they give rigid glassy solids at 77 K (liquid nitrogen temperature). These glasses can be handled by more conventional techniques than are needed for rare gas matrices; their use is further discussed at the end of chapter 4. Methane is sometimes used in the same way as the rare gases and nitrogen, and is included in table 2.2.

CO_2, SF_6 and CCl_4. Despite the even higher degrees of chemical reactivity and spectral interference to be expected from these larger molecules, they are used to some extent. They are particularly useful at rather higher temperatures than those necessary for the more inert matrix materials.

2.2 The structures of matrices

It is now pertinent to consider what the actual structure of the matrix, at the atomic or molecular level, may be. Three possibilities may be clearly defined, which we shall term the *single crystal*, the *glassy* and the *microcrystalline* models respectively.

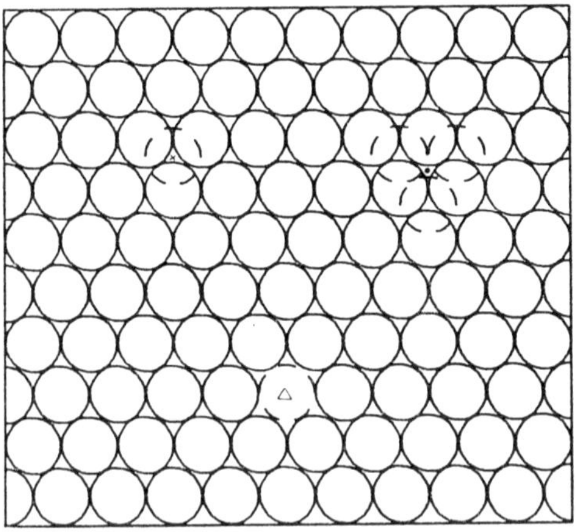

Fig. 2.1. Single crystal model.

Single crystal model. This has great advantages as a model, as the stable crystal structures of the matrix forming materials at low temperatures are well known, having been investigated by X-ray diffraction methods. In addition, the relative simplicity of this model, in which interstitial and substitutional sites are clearly defined (see fig. 2.1) makes it a favourite starting point for analysis of possible sites for matrix-isolated atoms and molecules.

Unfortunately, it is highly unlikely that a comparatively large volume of solid can be deposited in the form of a single crystal under the conditions used. These involve:

(i) a large-area condensation surface, so that many initial growth points must be expected;

(ii) fast freezing of a gas at a pressure and temperature considerably higher than the triple point to a solid at a temperature well below the triple point, so that the equilibrium is 'rushed through', and

(iii) the formation of the solid at such a temperature that diffusion of matrix atoms cannot occur in the bulk solid, preventing crystallisation after deposition.

Glassy model. In this case a random arrangement of matrix atoms is postulated (fig. 2.2); the density of the solid will be less than that of the perfect crystal, which is usually close-packed. The number of

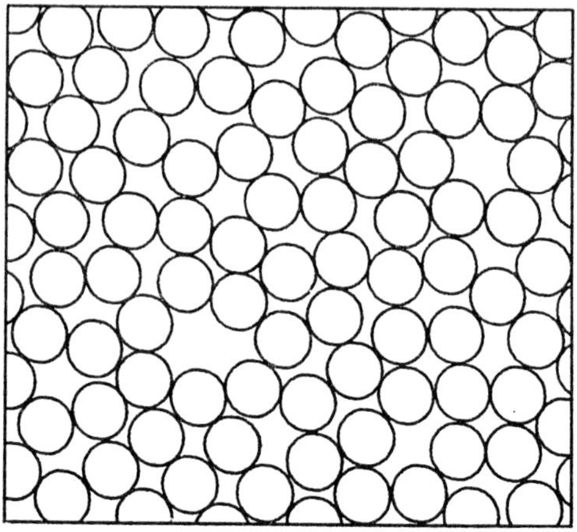

Fig. 2.2. Glassy model.

near neighbours for each atom will also vary randomly and will on average be less than 12, the number for a close-packed structure. As a result even 'single substitutional' sites, in which an isolated species replaces a single matrix atom, will vary in size and coordination, while the distinction between interstitial and substitutional sites will be to some extent lost.

The matrix-isolated species would then be expected to behave as if it were in an infinitely viscous liquid solution, and the effect of the matrix should be to broaden spectral bands. As we shall see later this is by no means commonly observed, and indeed one of the greatest advantages of the matrix isolation method is that spectral bands are usually extremely sharp, especially in the infrared and Raman. We must therefore reject the glassy model as appearing to predict broad bands where sharp bands are usually observed.

Microcrystalline model. The most plausible model is that of the microcrystalline solid. Here small regions of ordered structure exist, in which clearly-defined sites may be expected, but their crystal axes are not correlated and regions of random structure are required in between crystalline parts (fig. 2.3). These may be regarded as grain boundaries, so long as it is remembered that the crystalline regions may be so small that the 'crystalline' and 'boundary' regions take up comparable proportions of the solid.

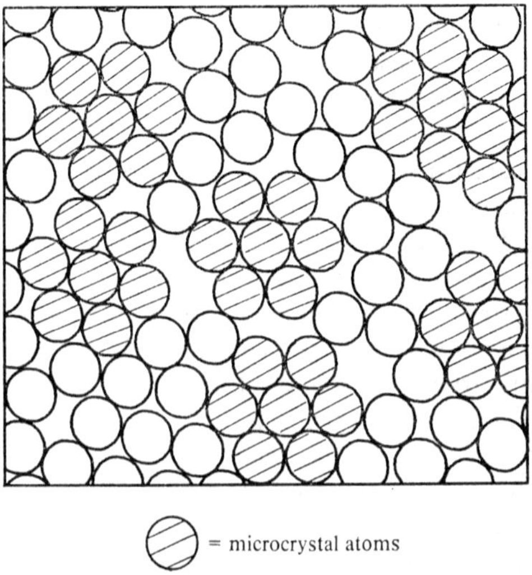

Fig. 2.3. Microcrystalline model.

At equilibrium it would be expected that any impurities (e.g. matrix-isolated species) would be segregated at the grain boundaries, but in the absence of diffusion this segregation cannot occur. When diffusion is allowed to occur it is certainly common for the sharp spectra characteristic of matrix-isolated species to be replaced gradually by broader bands more like those produced by substances in liquid solution.

This model is, however, very hard to treat quantitatively. We may not necessarily assume, for instance, that all crystalline regions adopt the structure that is most stable at the temperature of the experiment. In the absence of diffusion it is perfectly possible for two different crystalline phases to coexist, especially if they are physically separated by random regions. The range of possible sites in and between the crystalline and random regions is large, and it is extremely difficult to deal with the effects of long-range interactions between matrix-isolated species.

It is usual, therefore, for analysis of matrix spectra to begin by assuming that the single crystal model is appropriate. Other solid structures may then be considered if necessary as perturbations. We must thus discuss the *crystal* structures of typical matrix materials.

2.3 Crystal structures of matrix materials

The most commonly used matrix materials, the rare gases neon, argon, krypton and xenon, crystallise in the face-centred cubic structure (cubic close-packed). Each atom is surrounded by 12 equidistant nearest neighbours, and the symmetry of the site is that of the octahedron, O_h. This structure is the stable form for the solids at any temperature below the melting point.

There is, however, a less stable structure. This is simply the other close-packed structure, hexagonal close-packed. Again there are 12 nearest neighbours, but the symmetry of the site is lowered to D_{3h}. While this structure is less stable thermodynamically it is by no means certain that microcrystals with this structure may not form during deposition from the gas phase. Diffusion in the solid would lead to transformation to the more stable form, but is not always possible if the solid is rapidly cooled.

It has been shown that incorporation of even small amounts of nitrogen or oxygen into solid argon can render the hexagonal close-packed phase stable near the melting point, while a solid containing 40 % argon and 60 % nitrogen is stable with the hexagonal structure at all temperatures down to absolute zero. It is therefore rather important that the apparatus should be free from leaks during deposition of the matrix to prevent contamination of the solid with air, which would tend to alter the crystal structure.

It is apparent that other impurities, such as matrix-isolated species, in the matrix could have similar effects, at least locally. This could result in some microcrystalline regions having a hexagonal structure while others had a cubic structure. While X-ray diffraction can give information on the structure of a bulk solid it is not usually sensitive enough to detect small disturbances in structure in localised regions that may comprise less than 1 % of the total bulk.

Nitrogen, another important matrix material, has two stable solid phases, which transform reversibly into each other at 35.6 K in the presence of gaseous nitrogen. In the high temperature form (β-N_2) some degree of rotation of the molecules of nitrogen is possible, and they behave as near-spherical bodies in a hexagonal close-packed arrangement. Below 35.6 K, in the α-form, this rotation is 'frozen out' and a structure appropriate to cylindrical molecules is adopted. This is derived from the cubic close-packed structure. This is the only structure strictly relevant to matrix isolation, as solid nitrogen is rigid only below 20 K. However, the rearrangement involved in the phase-change implies that rapid cooling from the gas phase may result in a solid with molecules arranged in a less regular fashion than that found in α-N_2. The influence of impurities may well be able to make the β-form stable at lower temperatures, as for argon.

Carbon monoxide undergoes a similar phase-change in the solid, with the added complication that the dipoles of neighbouring molecules may not be trapped in the most stable mutual configuration when rotation ceases on cooling. This accounts for the 'residual entropy' of the crystalline solid at low temperatures, but does not seem to affect the results of matrix isolation experiments.

Solid oxygen has an even more complicated phase diagram, with three stable solid phases. Here the β–α transition occurs at 23.9 K, where the solid is quite rigid. It seems more than likely that 'supercooled β-O_2' could coexist with the stable α-O_2 in a microcrystalline matrix below the transition temperature.

2.4 Possible sites in close-packed lattices

We must now consider in more detail where in our hypothetical crystal lattice we may expect to find typical matrix-isolated species. Perhaps the simplest possible position is the *interstitial site* (see fig. 2.1) where the species is inserted between close-packed matrix atoms in an intact lattice. This is possible because close-packed spheres of equal size occupy only about 74 % of any volume, leaving 26 % unoccupied. In the cubic close-packed lattice there are two possible types of interstitial site, with four and six neighbouring atoms respectively. They are called tetrahedral and octahedral sites. The tetrahedral sites are extremely small, accommodating (without distortion) spheres of less than one-quarter the diameter of the close-packed spheres, and are probably not of significant importance. Even the octahedral sites are able to accommodate only spheres rather smaller than one-half the diameter of the close-packed spheres, and have only been convincingly shown to be occupied by hydrogen atoms (diameter 2.4 Å) in krypton and xenon matrices.

Table 2.1 gives the diameters of some atoms and ions; it will be seen that only monatomic positive ions seem likely to occupy interstitial sites on the grounds of size, but these are unlikely to be found isolated in matrices in any case. The occupation of interstitial sites by matrix-isolated species must be regarded as altogether exceptional, though diffusion of matrix-isolated species through the matrix presumably involves temporary occupation of such sites.

The other simple possibility is that of the *substitutional site*, in

TABLE 2.1. *Diameters of atoms, ions and molecules*

Typical atoms/Å								
					H	2.4		
	N	3.0	O	2.8	F	2·7	Ne	3.16
	P	3.8	S	3.7	Cl	3.6	Ar	3.75
	As	4.0	Se	4.0	Br	3.9	Kr	3.98
	Sb	4.4	Te	4.4	I	4.3	Xe	4.34
Typical ions/Å								
					H^-	4.0		
			Li^+	1.28	F^-	2.72		
			Na^+	1.98	Cl^-	3.62		
			K^+	2.66	Br^-	3.91		
			Rb^+	2.96	I^-	4.35		
Some typical molecules/Å								
			N_2	4.0	O_2	4.2		
			CH_4	4.5	CCl_4	7.1		

which one matrix atom or molecule is replaced in the lattice by the isolated species. Table 2.1 shows that many *atoms* may be expected to adopt such sites, being comparable in size with typical matrix atoms. The larger matrix atoms, such as krypton and xenon, may be expected to give rise to lattices capable of accommodating small molecules in substitutional sites. In such cases some distortion of the surrounding lattice may be expected to adapt to the non-spherical shape of the molecule.

It is clear, though, that with the exception of hydrogen, atoms in molecules will each require something approaching a single substitutional site in a rare gas matrix. Thus *molecules* will not in general be accommodated in single substitutional sites. It is more probable that we must consider matrix-isolated molecules as occupying *multiple substitutional sites*, in which several contiguous matrix atoms are replaced.

The removal of one atom from the close-packed lattice of the matrix leaves a single substitutional site with twelve nearest neighbours. Removal of two adjacent atoms leaves a site with 18 near neighbours; these are not of course evenly distributed, but serve to contain a more or less cylindrical space. The triple site formed by loss of three atoms in a triangle has 22 near neighbours, while the loss of three atoms in a row gives a site with 24 near neighbours. As more and more matrix atoms are replaced the possible shapes for the sites become more varied and the number of matrix atoms in what we may term the 'cage' will increase. We may expect that in general both the size and the shape of the matrix-isolated molecule will be reflected in the particular site adopted. The cage will not in most cases be ideally shaped, so that some degree of distortion of the lattice around the immediate cage must be expected.

Molecules larger than diatomics will, from the above analysis, have at least 20 matrix atoms in the surrounding cage of nearest neighbours. Large molecules will have many more near neighbours. The effects of small concentrations of impurities in the matrix may then be quite large, as a chance of finding an impurity molecule forming part of the cage increases with the number of atoms forming the cage. We shall return to this point below, but must first discuss some other important aspects of matrix materials, connected with the rigidity of the matrix and the mobility of species in it.

2.5 Rigidity and mobility

In a matrix isolation experiment it is important that no changes should take place while spectra are being recorded, and this implies that the matrix must remain rigid and prevent any diffusion of reactive species. A useful rule-of-thumb is that a matrix may be taken as rigid below 30 % of its melting point. Below this temperature essentially no rearrangement of matrix atoms or diffusion of isolated species is expected. In the range from 30 % to 50 % of the melting point the process of *annealing* of the matrix may occur. This is basically the rearrangement of the matrix structure at the atomic level towards the most stable crystal structure. Thus grain growth begins in this temperature range, while large trapped species will cause a local rearrangement to give the most stable possible cage. Small trapped species such as atoms and diatomics may be able to take part in the lattice rearrangement to such an extent that segregation at grain boundaries or reaction with neighbouring trapped species occurs.

Above about 50 % of the melting point the matrix must be regarded as non-rigid. *Diffusion* of trapped species now begins, and the ultimate effect is that all impurities will be segregated at grain boundaries and reactive species will be lost, as diffusion will occur until reaction results in formation of a large and stable species. This ultimate stage can rarely be achieved in practice because the vapour pressure of the matrix solid becomes too high and the matrix evaporates. Earlier stages in the diffusion process are more readily achieved, and the resulting effects are those to be expected from a general mobilisation of the solid. The smallest trapped species become mobile

TABLE 2.2. *Some significant temperatures for matrix materials*

Material	$0.3T_m$/K	$0.5T_m$/K	$T(P = 10^{-5}$ torr)/K	$T(P = 10^{-3}$ torr)/K
He			0.5	0.7
H_2	4.2	6.3	4.8	6.1
Ne	7.3	12.3	9.2	11.1
Ar	25	42	33	39
Kr	35	58	45	54
Xe	48	82	63	74
N_2	19	32	29	34
CO	20	34	33	38
O_2	16	27[a]	34	40
CH_4	27	45	40	48
CO_2	65	108	92	106
SF_6	67	111	73	87

[a] O_2 suffers a phase transition at 24 K.

at the lowest temperatures, and will diffuse until chemically trapped by reaction with some other reactive molecule. At higher temperatures larger molecules begin to diffuse.

Some pertinent data for typical matrix materials are given in table 2.2. We show the temperatures at which annealing and diffusion may be considered to begin ($0.3T_m$ and $0.5T_m$ respectively) and the temperatures at which the vapour pressure of the solid reaches 10^{-5} torr and 10^{-3} torr. These represent pressures that can be tolerated for long periods (of the order of hours) and very short periods (of the order of seconds) respectively. It can be seen that very precise control of temperature during diffusion is necessary if the matrix is not to be lost by evaporation.

Notable features among the data listed include the low vapour pressure of solid oxygen, which makes it possible to maintain the solid at $0.7T_m$ (40 K) for short periods without loss of the matrix. This means that diffusion processes can be followed essentially to completion in an oxygen matrix. The comparative rigidity of carbon dioxide and sulphur hexafluoride matrices, where diffusion does not begin until a temperature over 100 K is reached, should also be noted.

Oxygen, as noted earlier, suffers a phase-change at about 24 K. At this point we must expect a general mobilisation to occur although the temperature is below $0.5T_m$. The phase-changes in solid nitrogen and carbon monoxide occur at temperatures where the vapour pressure of the solid is high and are not important in controlled diffusion experiments.

2.6 Diffusion – mechanisms and results

As we have already implied, the diffusion behaviour of a trapped species will depend on its own nature as well as the nature and the temperature of the matrix. The temperature-dependence of diffusion rate is usually expressed in the familiar formula

$$\text{Rate} \propto \exp(-\Delta E/RT),$$

where ΔE is the activation energy for diffusion, expressed per mole of diffusing species. Unfortunately ΔE will vary with the size, shape and mass of the diffusing species, as it is related to the relative energy of substitutional and interstitial sites in the lattice, and it is not in general possible to calculate its precise value. Only empirical observations and generalisations about diffusion rates and mechanisms can therefore be made.

Thus it is observed that atoms and small diatomic molecules begin to show significant mobility in the annealing range, where the matrix is no longer completely rigid. Molecules with 3–7 atoms become mobile when general diffusion sets in, while large molecules remain until the lattice disintegrates around them.

It is easier on the whole to be precise about the results of diffusion than about the mechanisms involved. If our matrix-isolated species is reactive, diffusion will greatly increase the likelihood of it reacting with another matrix-isolated molecule. It is useful to distinguish two stages in such reactions, which are experimentally observed to occur separately in many cases. The first result of the close approach of two molecules during diffusion is the formation of *loose aggregates*, in which the spectroscopic properties of the interacting molecules remain distinct but are more or less perturbed by the influence of the neighbour. This sort of interaction may occur whatever the reactivity of the molecules and may be thought of as an essentially non-chemical effect. The second stage of the interaction occurs only if the two molecules are chemically reactive. In this case chemical reaction to form a *dimer* or other product occurs. The spectroscopic properties observed are then those of the new molecule, and those of the reactants are lost.

That these two stages of interaction are observed to be distinguishable suggests that the second, chemical reaction, stage requires either a reorientation of the aggregated molecules that cannot occur until the matrix becomes non-rigid or a small but appreciable activation energy. (These two requirements may in fact amount to the same thing.)

The process of aggregate formation need not, of course, stop when two molecules have come together. Other diffusing monomers may interact with the aggregate, and larger aggregates build up. If chemical reaction ensues these larger aggregates may form polymers, or a cluster of dimers, whichever is more stable. We may expect that aggregates and polymer molecules will in general be too large to diffuse themselves until the matrix has become very fluid.

Where the diffusing species is an atom or very small diatomic it is often found to be reactive enough to enter into chemical combination with trapped stable molecules whenever it encounters them, as well as combining with other reactive species. As in most experiments there will be more trapped stable molecules than atoms (which usually have to be generated in the matrix) the diffusing atom will be more likely to meet a stable molecule, and the major product will be that formed in this way rather than the dimer. Such small reactive species are also often found to be mobile during the annealing stage of matrix warm-up, so this type of reaction is characterised by the appearance of product peaks in the spectrum during annealing of the matrix. The intermediate formation of loose aggregates is not often observed in such cases.

Larger reactive species, which are mobile only when the diffusion temperature for the matrix is attained, may be expected to survive through the annealing stage and begin to show the effects of loose aggregation and then chemical reaction (if these are separately observable) at temperatures above $0.5T_m$.

Both annealing and diffusion, then, are important aids to the matrix isolation spectroscopist. Different species will be lost or formed at different temperatures, and it is possible not only to distinguish small mobile species from large species by their behaviour during annealing and diffusion but to establish which bands in a complex spectrum may or may not belong to any particular species. This is possible if we assume that the relative intensities of the various peaks due to a particular species will vary in the same way with its concentration. We can also distinguish between initial products that give bands whose intensity drops during diffusion (reactive products), products of the initial reaction that are not destroyed by diffusion (unreactive by-products) and substances formed by diffusion (secondary products).

Controlled annealing and diffusion are of great importance in the analysis of the results of an experiment. The ability to control the temperature of the matrix is therefore essential, and this is perhaps the most important facility of the newer microrefrigerators. The early matrix isolation experiments were done using liquid helium or hydrogen as refrigerants; the temperature could not be raised much above the boiling points of these liquids without physically removing the refrigerants. When this was done the temperature was liable to rise rapidly as no cooling was possible until the refrigerant was fed in again and the temperature dropped suddenly to 4 K or 27 K. A microrefrigerator makes it possible to control not only the temperature of the matrix but also the rate at which the temperature changes. The rate of heat extraction can be adjusted to be greater than, equal to or less than the rate of heat leakage into the matrix at any particular temperature and the experimenter is in control at all times. Much more precise study of the effects of annealing and diffusion is therefore possible.

2.7 Diffusion during deposition

Diffusion, unfortunately, is not only a controlled and useful tool, it is also an inherent property of the matrix system that will occur whenever conditions permit. It is particularly likely to be troublesome during deposition, when the matrix must be formed from the gas and conditions are changing so rapidly that little or no control is possible. It is obviously necessary that the condensation surface, on which the matrix is being deposited, is cooled to below the temperature at which the matrix becomes rigid, $0.3T_m$, but by no means sufficient. We must also ensure that the matrix attains this temperature as rapidly as possible after condensation. This implies that we must remove the heat released on cooling and freezing the matrix material fast enough to keep the sample already condensed below $0.3T_m$. This can in general be done only if the matrix is deposited slowly (no more than about 5 micromoles a second, or 18 millimoles (400 ml of gas at NTP) an hour).

Table 2.3 shows the cooling power in milliwatts required to cool and condense this quantity of each of the common matrix gases from 298 K to 20 K and to the rigidification temperature $0.3T_m$. Fortunately the cooling powers needed are well within the capacity of the microrefrigerators available today. The main problem is to ensure that the heat can be efficiently conducted away from the surface of the matrix, where deposition actually occurs, to the cooling surface and thence to the heat sink. Some technical details of these processes are given in the following chapter.

We still need to consider what happens in the actual deposition

TABLE 2.3. *Cooling requirements of matrix materials*

Material	Cooling power to 20 K /mW	Cooling power to $0.3T_m$ /mW
Ne	42	44
Ar	78	78
Kr	99	95
Xe	130	125
N_2	77	77
CO	81	81
O_2	85	85

region. It seems reasonable to suppose that, certainly for neon and probably for argon, nitrogen and carbon monoxide, a 'warm zone' will develop on the surface whose temperature will exceed $0.3T_m$ or even $0.5T_m$. In this zone we must expect annealing or diffusion to take place during deposition. Thus reactive species may be lost by aggregation or chemical reaction.

That this is so for some reactive species was shown by a study of the deposition of lithium atoms with rare gas matrices (co-condensation), where very few atoms escaped dimerisation during deposition, except in xenon. Clearly, in any experiment where reactive species are trapped from the gas phase this possibility must be borne in mind, and the deposition conditions controlled so as to minimise the probability of aggregation.

Favourable conditions in this respect will involve slow deposition, good thermal contact with the refrigerant, and a high ratio of matrix material to reactive species. Very long deposition times will then be necessary to ensure adequate concentrations of sample molecules, and it may only be possible to observe the strongest bands due to a species rather than all those expected. The sheer length of time required to deposit the matrix is now less of a limitation than in earlier years when the duration of an experiment was often limited by the available supply of liquid refrigerant. However, matrix isolation remains an activity involving 'anti-social' hours of work!

2.8 Pulsed matrix isolation

Non-reactive species and larger molecules in general will of course be far less affected by diffusion during deposition. It has recently been shown that for matrix isolation of stable molecules it is often

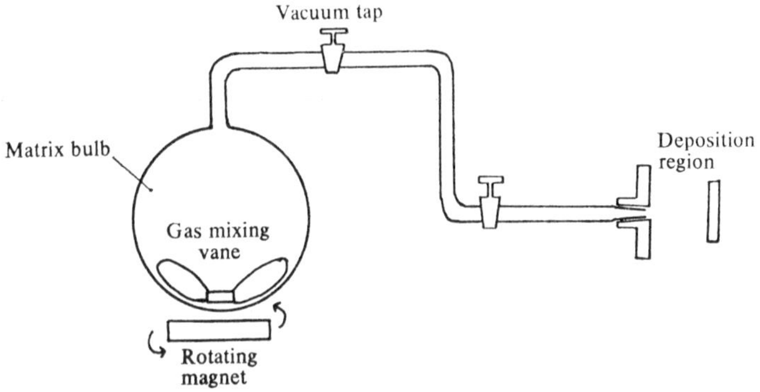

Fig. 2.4. Pulsed matrix studies.

advantageous to deposit quite large amounts of matrix in a single 'pulse' lasting a second or less. Each pulse is followed by an interval of several seconds to allow the matrix to cool again, after which a further pulse is deposited. The pulses are conveniently generated and controlled using a small volume of vacuum line between two taps (see fig. 2.4) which are opened and closed to fill the volume or to allow the matrix material to enter the deposition region.

In this way substantial quantities of matrix (and hence of matrix-isolated sample) can be deposited quickly. The matrix is often found to be optically clearer and to give sharper bands than a matrix formed by the conventional slow-deposition process. These effects are probably due to the thorough annealing of the matrix that must take place as the heat deposited with each pulse is conducted to the refrigerant. The pulse-deposition method has not yet been applied to direct deposition of matrices containing reactive species, and it does not seem likely that it will be generally useful in this respect.

2.9 Probability of isolation

The structure of matrix cages, which we discussed above, has some interesting implications for the matrix isolation spectroscopist. It is important to consider just *how* isolated the matrix-isolated species is, since if aggregation and chemical reaction lead to observable changes in the spectrum during annealing and diffusion similar interactions are bound to occur if two species are actually trapped in contact. This will effectively be the case if their respective cages overlap, so that one species forms part of the cage in which the other is trapped; for very reactive species non-nearest neighbour interactions may have to be taken into account as well.

TABLE 2.4. *Probability of isolation: single site occupation, weak interaction*

Matrix ratio	100	1000	10000
% non-isolated	11.4	1.2	0.1
% isolated	88.6	98.8	99.9

The probability of such interactions will depend on the *matrix ratio* (the ratio of matrix to trapped species, expressed in terms of atoms or molecules of each), on the particular type of site adopted and the effective size of the cage, and on the intensity of the interaction. We may take some examples that illustrate the sorts of behaviour expected.

Carbon monoxide is known to occupy a single substitutional site in most rare gas matrices; there are weak CO–CO interactions that are probably effective only between nearest neighbours. The probability of interaction is then simply the chance of finding a second CO molecule occupying one of the 12 sites that form the cage. The chance that this is *not* so is given by the formula $P = (1-r)^{12}$, where r is the reciprocal of the matrix ratio. For very small values of r the simple expression $P = 1-12r$ may be used. Table 2.4 gives the exact probabilities for several matrix ratios. We see that a matrix ratio of 1000 gives about 99 % isolation for a small molecule with weak interactions.

A larger trapped species has a larger cage and hence a larger probability of interaction. As an example, a species with diameter three times that of a matrix atom will occupy a site produced by the loss of a matrix atom and its original 12 neighbours; such a site will have a cage containing 122 atoms, so the probability of isolation is given by $P = (1-r)^{122}$, giving the probabilities shown in table 2.5. Here a matrix ratio of over 10000 is needed to ensure 99 % isolation. This calculation has ignored the fact that the presence of a second trapped species in the layer *beyond* the formal cage will also result in interaction; this fact requires the matrix ratios to be increased by a factor of ten or so over those shown in the table. It seems unlikely, on this basis, that such large molecules can ever have been studied in a truly isolated state.

A similar situation arises if species that interact strongly are involved. An example is afforded by atomic lithium, which is thought to dimerise if trapped within next-nearest neighbour distance (18 sites

TABLE 2.5. *Probability of isolation: multiple site occupation, weak interaction*

Matrix ratio	100	1000	10000
% non-isolated	70.7	11.5	1.2
% isolated	29.3	88.5	98.8

TABLE 2.6. *Probability of isolation: single site occupation, strong interaction*

Matrix ratio	100	1000	10000
% dimerised	16.5	1.8	0.2
% perturbed	41.3	6.4	0.7
% isolated	42.1	91.8	99.1

in all) and to interact significantly if found within about sixth-nearest neighbour distance (a further 68 sites). The probabilities of dimer or higher polymer formation, perturbation by long-range interactions and of effective isolation are then as shown in table 2.6. Again a matrix ratio of 10000 is needed to attain 99 % effective isolation.

The figures derived here become horribly significant when it is realised that until comparatively recently it was generally felt that a matrix ratio of 1000 was high and one of 10000 was ridiculous. Thus many early studies, where matrix ratios of a few hundred or less were commonly used, may have to be repeated, and their results treated with considerable suspicion until this has been done.

This is particularly true in that we have ignored the effects of diffusion during or after deposition, which can only serve to increase the likelihood of non-isolation in practice. It is found experimentally, for example, that carbon monoxide forms dimers or higher aggregates to the extent of several per cent at a matrix ratio of 1000 in argon, rather than the 1 % expected on the basis of our analysis. As mentioned earlier, lithium atoms appear to dimerise effectively completely even with matrix ratios of 10000 unless matrices that rigidify very rapidly (krypton or xenon) are used.

It is also clear that the matrix gas used must be free from impurities, even if these do not give rise to interfering spectral bands. The presence of, for instance, 1 % of nitrogen in argon would result in some 12 % of all sites having a cage containing at least one nitrogen molecule even if only single substitutional sites are considered. This sort of effect may well be responsible for some of the splittings reported in matrix spectra (see chapter 6). More reactive impurities such as oxygen or water will produce even more disastrous effects, as a significant proportion of any reactive species may well be removed by reaction with impurities in the cage – without the need for any diffusion.

2.10 Spectral compatibility

Before we go on to discuss the technical details of matrix production we must consider one further important physical property of the matrix in more detail. This is simply the absorption, emission, Raman or e.s.r. spectrum, which might interfere with our spectroscopic study of the matrix-isolated sample.

Rare gases. The rare gases are very nearly ideal in this respect. Having no unpaired electrons they give no e.s.r. spectrum, though the nuclear magnetic moment of some isotopes of krypton and xenon gives rise to splittings in the e.s.r. spectra of some species isolated in these matrices. They are monatomic and have no molecular vibrations to give rise to infrared and Raman bands. Their high ionisation potentials mean that the electronic absorption spectrum does not begin until rather high energies (short wavelengths). The first strong absorption corresponds to the first resonance transition $(n+1)$s–np. The wavelengths associated with these transitions are shown in table 2.7. The solid matrix will have weaker absorptions at somewhat lower energy, and these matrix materials will therefore become effectively opaque at rather longer wavelengths than those listed here. The exact 'cut-off' point will depend on the sample thickness.

From these short wavelength limits the rare gas matrices are free from interfering absorption bands throughout the ultraviolet, visible and infrared regions. No emission will be excited unless the sample is irradiated with light of wavelength shorter than the cut-off point, and no Raman transitions are possible. Only in the far infrared region, and the corresponding portion of the Raman spectrum, may there possibly be bands associated with the motions of the crystal lattice (lattice modes). Such modes have not been observed in matrix isolation studies. Furthermore, a study of pure crystalline argon in the far infrared showed no discrete bands between 100 cm^{-1} and 20 cm^{-1} even with 25 cm thickness of solid. For all practical purposes,

TABLE 2.7. *First resonance transitions of rare gases*

Rare gas	Wavelength/nm
Ne	80
Ar	104
Kr	125
Xe	160

TABLE 2.8. *Lower wavelength limits for molecular matrices*

Molecule	Wavelength/nm
CH_4	140
N_2	200
CO	205
CO_2	220
SF_6	220
O_2	1250

then, the rare gas matrix may be considered free of bands in absorption, emission, Raman and e.s.r. spectra except for the short wavelength (vacuum ultraviolet) region.

Molecular matrix materials. These are much more likely to give interfering spectral bands than are the rare gases. There are now vibrational modes of the molecule as well as of the lattice, and in most cases electronic transitions at lower energy than those of the rare gases. The lower wavelength limits for the molecular solids commonly used as matrix materials are listed in table 2.8. It will be seen that only methane is usable below 200 nm, while oxygen absorbs significantly over the ultraviolet, visible and near infrared regions.

Oxygen, alone among the common matrix materials, has unpaired electrons. It therefore gives rise to e.s.r. signals and is not usable for e.s.r. studies of isolated materials.

The internal vibration frequencies of the molecular substances are listed in table 2.9, together with an indication of the infrared or Raman activity of each mode. The frequencies listed are perhaps misleadingly precise; the large quantity of matrix material inevitably present will in most cases result in broad bands of zero transmittance in the infrared and broad strong bands in the Raman. In some cases overtone and combination bands will also be strong enough to cause serious gaps in the observable infrared spectrum.

TABLE 2.9. *Internal vibrations of molecules*

Molecule	Frequency/cm^{-1}	Activity[a]
N_2	2330	R
CO	2140	I, R
O_2	1560	R
CO_2	2340	I
	1388	R
	1277	R
	667	I
CH_4	3019	I, R
	2916	R
	1533	R
	1306	I, R
SF_6	940	I
	770	R
	640	R
	610	I
	520	R

[a] I = infrared active; R = Raman active.

TABLE 2.10. *Lattice modes of molecular crystals*

Molecule	Frequency/cm^{-1}	Activity
N_2	83	R
	69	I
	49	I
	37	R
	33	R
CO	86	I
	50	I
	47	R
CO_2	130	R
	114	I
	90	R
	73	R
	68	I

The lattice modes of molecular solids are more likely to be a problem than are those of the rare gas solids because of the greater complexity of the crystal structures. They have been observed in the infrared and Raman spectra of solids such as pure crystalline nitrogen, carbon monoxide and carbon dioxide; the data are listed in table 2.10.

Such bands have not so far been a serious problem as the corresponding spectral region has not been much studied. It seems likely that the microcrystalline nature of the solid matrix will result in such bands being weaker and broader than in pure crystalline solids.

It has been suggested that combinations of lattice modes with internal vibrations of trapped molecules may be involved in the generation of 'matrix splittings'. This topic is discussed at greater length in chapter 6.

3 Cryogenic, vacuum and oven technology

As we have seen in chapter 2, it is necessary to maintain a rare gas matrix at a low temperature to prevent diffusion of reactive species and their destruction by aggregation. We must now consider in more detail some of the technology involved in this. The maintenance of very low temperatures also implies the use of vacuum techniques to prevent heat leakage by gas conduction, convection and condensation, so we must also include some aspects of vacuum technology. Finally, many matrix experiments somewhat paradoxically involve the use of ovens to evaporate monomeric species from normally polymeric solids or to generate them by high temperature reactions. We shall comment on the design of such ovens and the special problems that arise in the combination of high temperature and very low temperature apparatus.

3.1 Cryogenic technology

Dewar vessels. The primary necessity for the use of vacuum in cryogenics, as mentioned above, is due to the insulating capacity of a vacuum, where the conduction and convection of heat possible in a gas are eliminated, and only radiation remains as a means of heat transfer. We are all familiar with the Dewar flask, where a double-walled glass vessel is used so that the space between the walls may be evacuated. This provides excellent insulation, which is further enhanced by silvering the internal walls of the evacuated cavity to reduce the efficiency of radiation and absorption of heat. Such a simple glass flask may be used to store liquid nitrogen at 77 K for hours at a time. Similar vessels with thin metal walls (for greater strength, see fig. 3.1) are used for longer-term storage of large quantities of liquid nitrogen and for its transport by road, rail or air.

Storage of liquid helium and hydrogen. Simple Dewar vessels are inadequate for the storage of refrigerants such as liquid helium and liquid hydrogen, which are needed for matrix isolation using rare gas matrices. There are several reasons for this, the most important of which are:

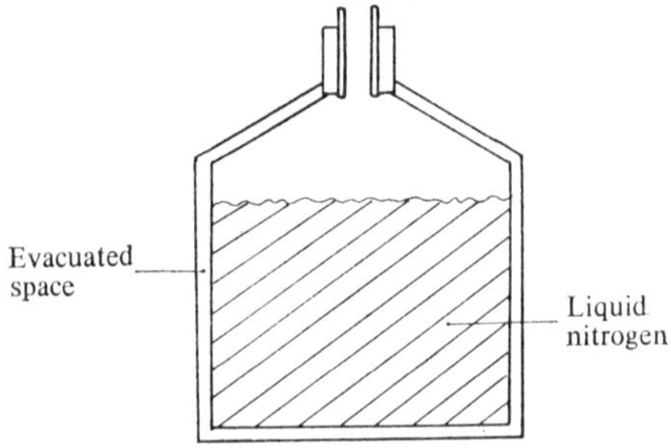

Fig. 3.1. Simple Dewar vessel.

(i) The rate of heat transfer by radiation is proportional to $(T_a^4 - T_b^4)$, so that more heat is transferred to the inside wall as its temperature drops;

(ii) The latent heat of evaporation of a liquid falls as its boiling point drops, so a given amount of heat evaporates far more liquid helium than liquid nitrogen;

(iii) The temperature of liquid helium or hydrogen is far below the freezing point of air or nitrogen, so unless some precaution is taken to prevent contact between the evaporating refrigerant and the atmosphere a solid plug of frozen air rapidly blocks the neck of the flask, preventing escape of refrigerant gas and leading eventually to an explosion;

(iv) The very low temperatures also mean that the vapour pressures of nitrogen, oxygen and indeed all substances except hydrogen and helium are extremely low, so that any surface cooled by liquid hydrogen or helium will condense any residual atmosphere. The latent heat of evaporation is released in this condensation, and this causes the loss of a much larger quantity of refrigerant (see (ii) above). Unless the insulating vacuum is effectively perfect or is continually pumped a continuous heat supply is afforded by such condensation of air from microscopic leaks. In addition, the build-up of a film of condensed air rapidly lowers the reflecting power of the silvered wall and radiative heat transfer also increases.

Containers for liquid hydrogen and helium. These physical factors must all be overcome before liquid hydrogen or helium can be stored or used effectively. It is not practicable to do this in general using glass apparatus, because of the internal strains set up by cooling and heating and the inevitable complexity of the apparatus. Metal apparatus is universally used, despite the higher heat conductivity of most metals; the high strength of the material means that much thinner walls can be used, thus reducing the rate of heat conduction *along* the wall. The metals most often used are selected for mechanical strength, ease of fabrication and, where necessary, low heat conductivity. Brass and copper are used where possible, stainless steel where the minimum heat conductivity is needed.

The problem of *radiative heat transfer* is greatly diminished if the inner apparatus is surrounded by an outer sheath held at 77 K, liquid nitrogen temperature. In principle, this simple device lowers the rate of radiation to the inner container by a factor of about 200.

In storage vessels this is simply achieved using an outer metal Dewar flask containing liquid nitrogen; the liquid helium or hydrogen occupies a smaller metal flask inside this, and the space between the walls is evacuated (see fig. 3.2). If an apparatus contains extensions that must remain at very low temperature but cannot conveniently be surrounded by a nitrogen Dewar, metal sheaths are used to 'shroud' the inner container. These are cooled by conduction from a reservoir of liquid nitrogen or other part of the apparatus held at or near 77 K (see fig. 3.3).

The problem of *air plug formation* is countered by the use of a hydrogen or helium atmosphere above the refrigerant. This atmosphere is connected to a large 'ballast volume' containing the same gas to contain a pressure surge such as will inevitably follow any substantial heat leak into the liquid refrigerant.

The condensation of air on to the cold wall of a cryogenic container (often called *cryopumping*) can in principle be prevented only by working at a pressure below the vapour pressure of air at the wall temperature. At 4.2 K (liquid helium) the vapour pressure of nitrogen is about 10^{-87} mmHg, which is not attainable by normal means. The pressure is limited in practice to a value such that the rate of condensation is negligible. For instance, in an apparatus with a total volume of 5 litres, a pressure of 10^{-6} mmHg implies a total gas content of 2.6×10^{-10} moles, or about 7 nanograms of nitrogen. The heat released on condensation of this quantity of gas is negligible (about 2 microjoules). Evacuation to about 10^{-6} mmHg and the elimination of leaks significant at this level is thus adequate. At the same time this also ensures the absence of significant nitrogen impurity in the matrix (deposited usually at several micromoles a second).

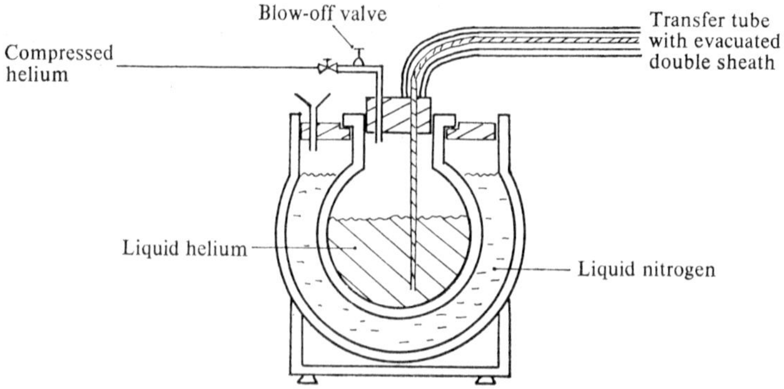

Fig. 3.2. Storage of liquid helium.

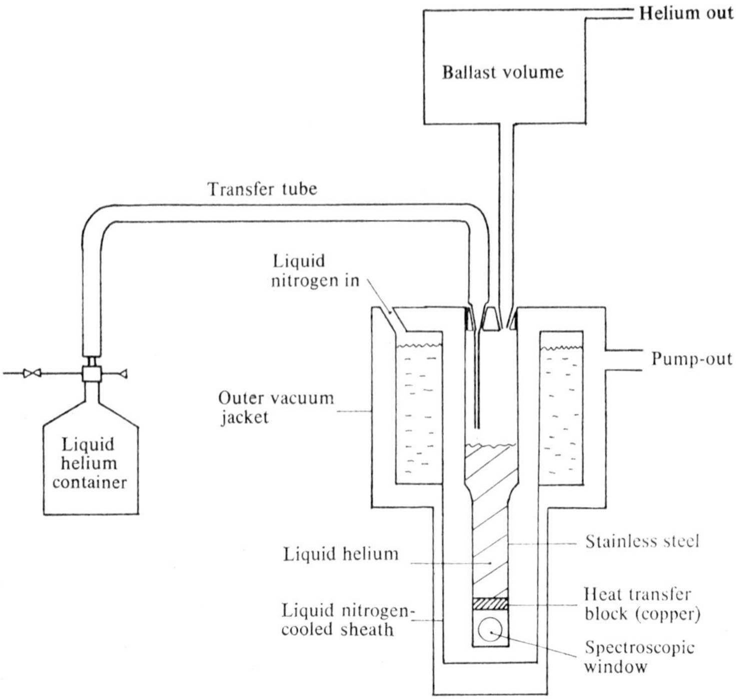

Fig. 3.3. An apparatus cooled by liquid helium.

Of course, the problem of leaks is less serious if the Dewar vacuum can be pumped continuously, and this is done where possible. Storage vessels, which must be easily moved, cannot be pumped continuously, and are consequently manufactured to very rigorous standards and passed for use only after a prolonged practical test of evaporation rate of refrigerant. This test is far more sensitive than any other method for detecting leaks.

3.2 Vacuum insulation in matrix apparatus

The use of vacuum insulation in cryogenic apparatus makes it relatively easy to build up inert gas matrices by condensation on a cooled surface. The matrix gas is simply fed into the vacuum vessel through a nozzle pointing at the cooled surface. Some of the gas is condensed, and any that escapes immediate condensation is removed either by the pumps or by the 'cryopumping' action of other cold parts of the apparatus. The rate at which gas is introduced must be carefully controlled so that excessive pressure rise does not occur, otherwise heat transfer across the Dewar vacuum becomes too great and the refrigerant evaporates.

Unfortunately, this system leaves the condensing surface exposed to any slight air leaks in the vacuum system, and to any substances that may evaporate into the system, such as water from metal or glass surfaces and carbon monoxide from ovens. The matrix isolation spectroscopist must be constantly on his guard against any contamination of the matrix with such materials, which will react with many of the species he is interested in, and may perturb the spectra of others. It has repeatedly been observed, for instance, that when lithium is evaporated and condensed into argon matrices the major products are oxides and hydroxides derived from reaction of oxygen or water with lithium atoms before, during and after deposition.

Air leaks can, in principle, be discovered and sealed. Evaporation from surfaces at or near room temperature is more of a problem; it can be overcome only by the most scrupulous cleaning of all metal and glass surfaces before evacuation, and by prolonged pumping before the apparatus is cooled down. This pumping period can be shortened if the apparatus can be heated, or 'baked', but this is rarely practicable. Evaporation from hot surfaces in and near ovens is also difficult to prevent; the best method of preventing interference by such effects is to 'bake' the oven at temperatures slightly lower than those to be used in the experiment for some hours under a vacuum before cooling down the condensation surface. Any baking carried out after cooling down is more-or-less useless, as condensation on the cold window is inevitable unless a movable shutter is arranged to block the 'line-of-sight' path from oven to window.

Cooling a practical matrix isolation apparatus. The principle involved in cooling any apparatus to 4 K with liquid helium or to 20 K with liquid hydrogen is simple, and we have considered some of the precautions needed in the design of the apparatus. We must now go more deeply into some of the practical problems that arise in matrix isolation experiments.

The earliest method used to cool the apparatus was to *transfer* liquid helium or hydrogen from a 'double Dewar' storage container to a refrigerant vessel whose base was in good thermal contact with the surface to be cooled. This transfer operation is less easy than it sounds; the refrigerant must be protected from heat leaks and from contact with air at all stages.

The refrigerant is driven from the storage vessel using a pressure of refrigerant gas (helium or hydrogen), along a tube protected by a double evacuated shroud, into the coolant container (see fig. 3.3). This is surrounded by a liquid nitrogen-cooled shroud. The coolant container must first be pre-cooled with liquid nitrogen, to reduce the amount of refrigerant needed to cool it to the operating temperature; the liquid and gaseous nitrogen must then be removed by flushing with helium or hydrogen.

The refrigerant gas evaporated during further cooling of the container is vented, not to the atmosphere, but to a ballast volume, which serves to contain pressure surges, and finally either collected by a compressor or vented outside the building. Helium, which is safe but expensive, is often collected and reliquefied, while hydrogen, which is cheap but potentially an explosion hazard, is usually allowed to escape.

Practical considerations prevent the use of large refrigerant containers in spectroscopic apparatus, so the experiment must either be completed in a few hours before the refrigerant evaporates completely, or else the transfer process must be repeated at intervals to refill the container. The transfer tube is usually lifted clear of the liquid surface during an experiment to reduce losses due to conduction of heat down the tube, but is not removed completely from the apparatus. It may be supported from a pulley system with a counterweight so that it may be raised and lowered as needed.

The explosion hazards associated with the use of liquid hydrogen in an oxygen-containing atmosphere are so great that additional shielding is provided all round the apparatus, which is thus effectively contained in an atmosphere of hydrogen. A larger ballast volume is thus available to contain pressure surges; it is provided with a large one-way vent (a weighted flap) to the outside of the building so that no internal accident is likely to release large quantities of hydrogen into the laboratory. In addition a 'hydrogen alarm', operated by the heating of a platinum coil if hydrogen is present in the atmosphere, is often mounted above the apparatus to detect minor leaks. Relatively few laboratories have used liquid hydrogen as a refrigerant, despite its cheapness, because of the dangers involved.

3.3 Microrefrigerators

The expense and inconvenience associated with the use of liquid helium and the hazards involved in the use of liquid hydrogen, have naturally led to the development of alternative methods of attaining and maintaining low temperatures. There are now commercially available *microrefrigerators* of various types that can not only provide steady temperatures near 77 K, 20 K and 4 K by the use of nitrogen, hydrogen and helium respectively as working fluids, but also enable in each case any higher temperature to be maintained steadily by a simple change in the rate of heat extraction. This has the great advantage of permitting controlled annealing and diffusion to be carried out in matrices.

There are two basic types of microrefrigerator currently available for this kind of application, depending on the Joule–Thomson effect and the Solvay cycle respectively.

Joule–Thomson type. This type uses the cooling of a gas expanding through a nozzle; the cool gas is used to pre-cool further compressed gas before expansion, while the expanded gas may be vented to the atmosphere or recompressed and recycled after cooling. This is the basis of the Linde process for production of liquid air. Unfortunately, each gas has a characteristic Joule–Thomson inversion temperature *above* which the effect is inverted, so that expansion through a nozzle causes a *heating* of the gas.

For nitrogen and air this inversion temperature is above room temperature, so that the simple process may be used to liquefy nitrogen or air. For hydrogen the inversion temperature is 193 K, and the process does not work efficiently unless the gas is cooled substantially below this temperature before expansion. This is most

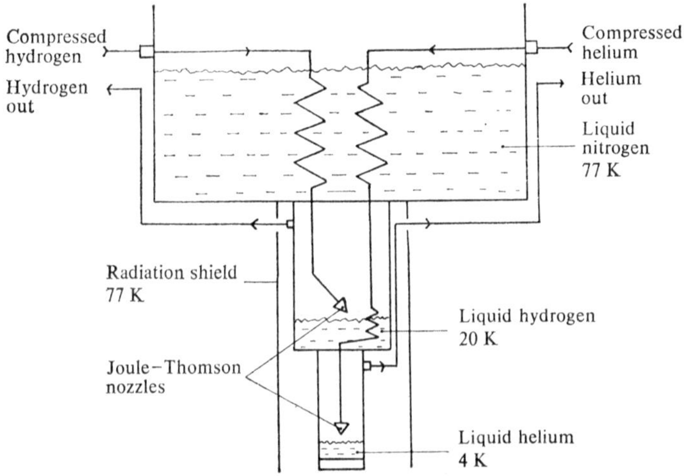

Fig. 3.4. Production of liquid helium by a two-stage Joule–Thomson process.

conveniently done using liquid nitrogen at 77 K as a pre-coolant; liquid hydrogen is then produced by recycling the pre-cooled hydrogen. A two-stage process in which nitrogen is liquefied first and used to pre-cool hydrogen enables a hydrogen liquefier to work independently of outside supplies of liquid nitrogen, but it is usually simpler and cheaper to supply the liquid nitrogen pre-coolant from outside the apparatus, which then becomes considerably smaller and less complex.

For helium the inversion temperature is 100 K, and even pre-cooling in liquid nitrogen is not enough to start an efficient cycle. A two-stage process, in which hydrogen, pre-cooled in liquid nitrogen, is liquefied and used to pre-cool helium, is needed (see fig. 3.4). The equipment required is understandably elaborate, but can be used to maintain equipment continuously at liquid helium temperatures for long periods. These microrefrigerators are normally run as open-ended systems to reduce the complexity of the equipment; this means that considerable amounts of compressed (cylinder) hydrogen and helium are needed and must be vented safely or recovered. Their great advantages over 'liquid transfer' systems include:

(i) the absence of large volumes of liquid hydrogen and helium – only fine sprays of droplets need to be produced to provide constant cooling;

(ii) the *rate* of cooling can be varied by controlling the flow rate of gas through the system, so temperature control is possible by balancing heat input to the sample against cooling rate;

(iii) the temperature may also be controlled to a smaller extent by raising or lowering the pressure in the expansion volume, thus altering the temperature of the liquid in equilibrium with the gas;

(iv) compressed gases are cheaper, more readily available and safer than the liquids, and can be stored indefinitely.

Solvay cycle type. The second type of microrefrigerator has been developed more recently. It uses the principle that in expanding against a *constant* pressure a gas does work, so that cooling *must* occur whether the gas is above or below the Joule–Thomson inversion temperature. Thus the apparatus can operate with any gas, at almost any temperature. The lower limit to the temperature is, of course, the boiling point of the working fluid, at which point the gas liquefies and cannot be used as working fluid in the same way. (Domestic refrigerators use a different cycle in which the working fluid is sometimes gaseous, sometimes liquid, but the two sorts of cycle cannot usefully be carried out in the same apparatus.) As the rate of heat transfer depends on the density (and hence the pressure) of the working fluid it is necessary to run the apparatus at rather high pressures to obtain adequate cooling power. The relevant boiling point is thus rather higher than that appropriate to atmospheric pressure.

The apparatus (see fig. 3.5) could in principle be used in open-ended or closed-cycle forms. In the open-ended system the gas enters at high pressure, being cooled by the cold apparatus and displacing a plunger upwards. Gas at the top, 'warm' end is displaced through a fine hole into a surge volume. When the inlet valve closes the high pressure gas at the 'cold' end expands until the pressure at warm and cold ends equalise. In this process the high pressure gas is cooled, as it does work, while the gas at the warm end warms up as work is done on it. An exhaust valve then opens and gas from the cold end leaves the apparatus, and the plunger moves downwards as gas returns from the surge volume to equalise the pressure. The escaping gas from the cold end cools the apparatus (plunger and warm end) as it passes. The cycle is then repeated by the introduction of more compressed gas. Thermodynamically, the work required to pump heat from the cold end to the warm end is provided by the expansion of compressed gas. Typical working pressures are 20 atm at the high pressure end, reducing to 6 atm during the working expansion.

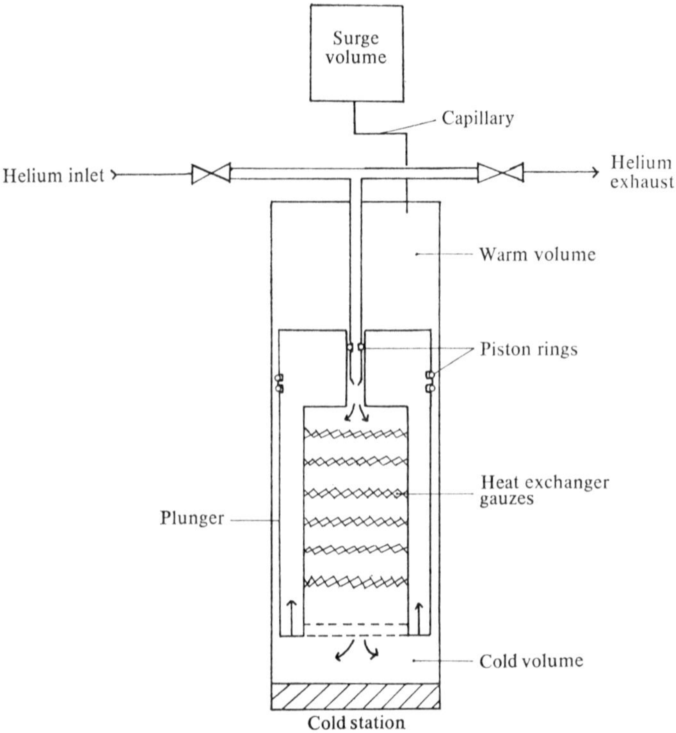

Fig. 3.5. Single stage Solvay refrigerator.

This form of the apparatus is extremely simple and uses very little electric power (only for switching of valves) and no liquid nitrogen pre-coolant. If operated with nitrogen as working fluid a single stage system will operate down to about 100 K. A two-stage system (where heat is pumped in two steps, from very cold to cold and from cold to warm) working with helium will operate down to about 10 K. It still suffers from the disadvantage that large quantities of compressed gas are needed.

This is overcome quite easily, as only one working fluid is involved for any one apparatus. The exhaust gas is recompressed to the working pressure (20 atm) and returned, after cooling to room temperature, to the beginning of the cycle. This converts the apparatus into a closed-cycle system, in which the work needed is supplied as electrical power to the compressor. No helium is lost, and the process is just as efficient as in the open-ended form. Such an apparatus may be kept running for days at a time, greatly extending the time-scale of matrix isolation experiments. Again, temperature control is easily achieved by reducing the pressure differential in the apparatus (and hence the work done in each cycle) using a 'controlled leak' by-passing the compressor.

Microrefrigerators and matrices. Microrefrigerators are now available in the form of cylinders which may readily be substituted for the coolant containers of matrix isolation systems. The weight of the helium models is considerable, so they are mounted in heavy metal jackets that also serve as the insulating vacuum vessel. A typical apparatus is shown in fig. 3.6. Spectroscopic windows are provided in the jacket, together with a spray-on nozzle for introduction of matrix material, and ovens or photolysis windows may be added if needed. If the microrefrigerator unit is mounted in the outer jacket with a double O-ring seal it may conveniently be rotated so that the inner, cooled window is aligned with the spectrometer or with the spray-on tube or the photolysis window. The very cold portions of the apparatus are conveniently shielded from radiation by a heat shield cooled by the intermediate cold stage of the two-stage microrefrigerator.

3.4 Temperature measurement and control

It is important for the matrix isolation spectroscopist to know what the temperature of his sample is, as only then can he know what degree of annealing or diffusion may be occurring, or indeed achieve reproducible experimental conditions. While there are a number of devices that could in principle be used (for example vapour pressure thermometers using helium or hydrogen, resistance thermometers, etc.) the thermocouple is by far the most useful. The particular type used depends on the range and precision of the results needed, but a chromel/0.07 % iron in gold combination is often used. The reference junction may be immersed in liquid nitrogen (77 K) or in ice-water (273 K).

The transfer of heat from the sample on its window or plate to the refrigerant must usually be made as efficient as possible, and careful design of the 'heat-transfer' section is essential. It usually consists of a solid copper block, the top of which forms the base of the refrigerant container or the 'very cold' end of a microrefrigerator (see fig. 3.7). A spectroscopic plate may be held firmly against the block by a metal plate held on with screws. The use of a gasket of

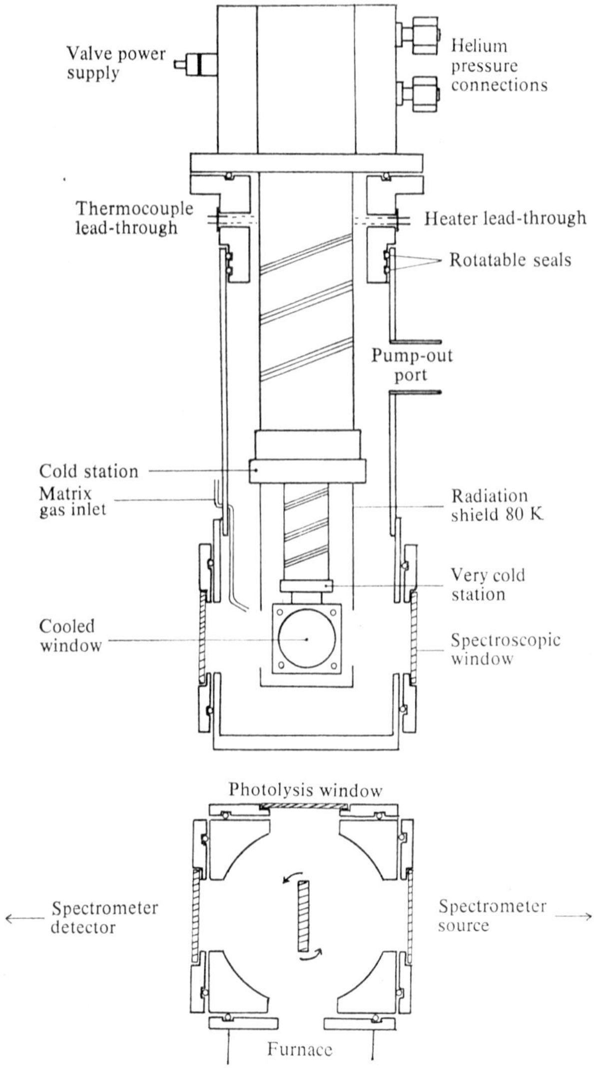

Fig. 3.6. A complete apparatus for matrix studies.

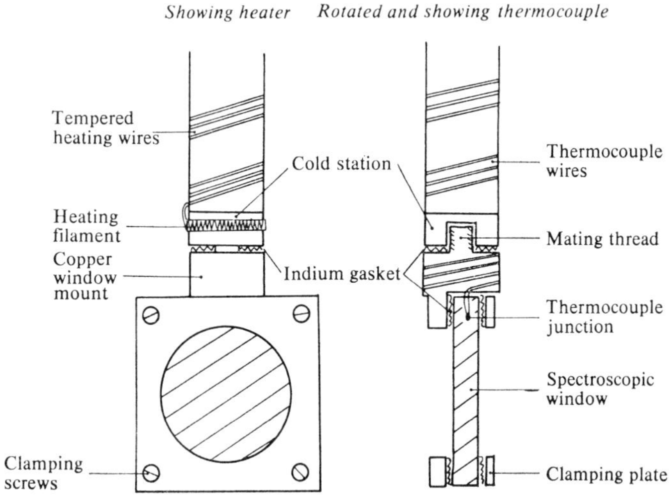

Fig. 3.7. Window mounting and thermocouple assemblies.

indium, which remains soft at very low temperatures and has a reasonable thermal conductivity, prevents the differential contraction of the metal and the window from cracking the latter.

Heat absorbed by the sample or the window (such as latent heat released during condensation, radiation absorbed during photolysis or spectroscopic study, and stray heat leaks) must then be conducted to and through the window, the gasket and the copper block. The last step is usually the most efficient, the first the least. It is important, then, that the temperature monitored by the thermocouple is that of the window and not that of the copper block. The thermocouple junction must then be held firmly in contact with the window, preferably in a small hole drilled in it, and must not touch any metal part in good thermal contact with the copper block or the refrigerant. On the other hand, to reduce the possibility of a significant heat leak to the window through the thermocouple wires they are 'tempered', or wound around the cooling section of the apparatus so that their temperature is reduced from that of the outside before they reach the window. (The temperature of the wires is irrelevant to the reading given; only the temperatures of the two junctions are important.) For the same reason the wires of the thermocouple are made as thin as possible to reduce heat leaks by this route. The e.m.f. produced by the thermocouple should be measured using a 'bridge' arrangement that does not result in any current flowing in the junctions that might give rise to resistive and thermoelectric heating effects. Ideally, any thermocouple circuit should contain only two junctions; this is not possible in practice, but any additional junctions between dissimilar metals (as for example those to the bridge or other measuring system, and those at the point where the wires enter the vacuum shroud) must be kept at a uniform temperature to prevent errors in the temperature measurement.

Temperature control is often achieved, as noted above, simply by adjusting the rate of cooling until it is less than the rate at which heat is entering the sample, so that the temperature rises until a new equilibrium temperature is reached. If the temperature rise attainable by this method is insufficient the rate of heat leakage may be artificially increased using a small heating coil on the heat transfer block. Where bulk liquid refrigerants are used this is indeed the only way the temperature of the sample can be raised much above the boiling point of the refrigerant. Smaller temperature rises can be controlled in this case by increasing the 'boil off' pressure on the refrigerant, and thus its boiling point.

3.5 Vacuum techniques

We have seen that the maintenance of low temperatures and the formation of solid matrices from the rare gases require the use of a vacuum for insulation and for protection of the matrix from contamination. The particular techniques used in any vacuum apparatus are conditioned by the scale of the apparatus and the pressure needed. For matrix work it is almost always necessary to have a two-stage pumping system with a liquid nitrogen-cooled trap to give an ultimate pressure of about 10^{-6} mmHg, for reasons explained above. The most common arrangement is that of a mechanical (rotary) pump, which can provide an ultimate pressure of the order of 10^{-3} mmHg backing an oil-diffusion pump to reduce the ultimate pressure to 10^{-6} mmHg and to increase the speed of pumping near the lower pressure limit of the rotary pump. The rotary and diffusion pumps may each be one-, two- or three-stage in their action depending on the pumping speeds needed.

In any vacuum system, as opposed to a single pump, the ultimate pressure attainable and the speed at which it is reached are greatly dependent on the precise geometry of the system. Any constriction or bend will markedly reduce the effective pumping speed at pressures where 'molecular flow' occurs, as most molecules will need to make at least one wall-collision to pass the obstruction. It is most advisable to arrange that low pressure gases move along line-of-sight paths whenever possible.

As most matrix isolation systems are mounted on spectrometers it is usually necessary to pump out more or less horizontally from the outer vacuum jacket to avoid the spectrometer. Most diffusion pumps, on the other hand, are arranged to pump vertically downwards, as this simplifies oil-return from the condenser to the boiler. There must then be at least one 90° change in direction between apparatus and pump and there should be no more if at all possible. A valve mounted between the apparatus and the pump, if any, should be designed to constrict the flow of gas as little as possible when open; a 'butterfly' valve is most suitable, though other designs based on the closing of a large orifice by a flap would be adequate.

The cold-trap, designed both to prevent back-diffusion of vapours from the pumps to the very cold sample mount and to protect the pumps from reactive vapours released when the matrix evaporates, should also be of a 'flow-through' type with no abrupt changes of direction or constrictions. Finally, it is important that the entire pump-out tube connecting the apparatus with the pumps be as short and as wide as possible, and that any couplings, flexible sections and so on should not act as constrictions.

On the high pressure side of the diffusion pump the gas flow is much less affected by bends and constrictions. Small-bore copper pipe is often used for connections in this part of the pumping system.

Pressure measurements. It is necessary, of course, to have some means of monitoring pressure at various points in the system. Two relatively inexpensive types of pressure gauge are useful in our context.

Pirani gauge

This operates best over the range 10^{-1}–10^{-3} mmHg and is used to monitor the backing pressure maintained by the rotary pump. Its operation depends on the conduction of heat, from a filament carrying an electric current, through the surrounding gas. Changes in pressure change the rate of heat conduction, hence the equilibrium temperature of the filament and thus its resistance. The filament forms one arm of a bridge circuit and changes in resistance are registered on a meter.

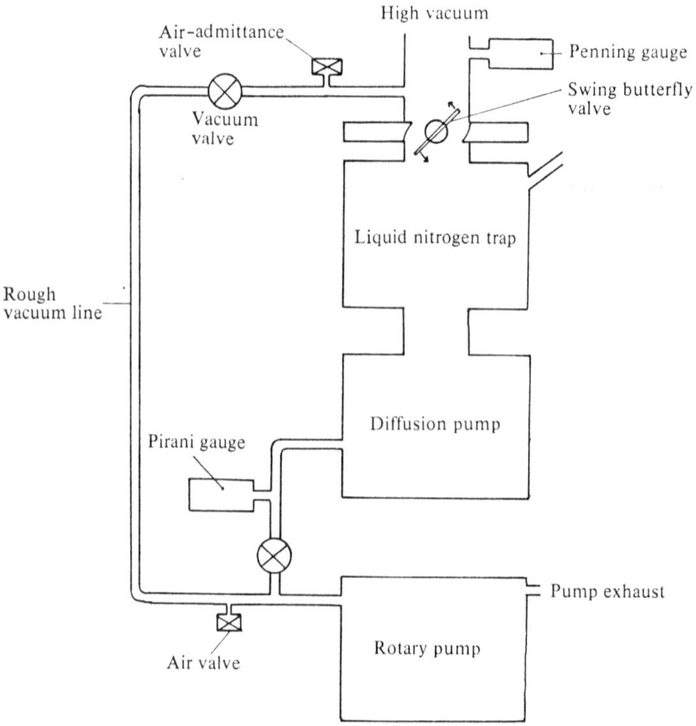

Fig. 3.8. Schematic vacuum system.

Penning gauge

This operates in the range 10^{-2}–10^{-7} mmHg, measuring the current flowing in a high voltage discharge in the gas whose pressure is to be measured. The effective resistance of the discharge cell varies with pressure and with the nature of the gas. Penning gauges are used to monitor the pressure on the low pressure side of the diffusion pump and in the matrix isolation apparatus itself.

A diagram of a complete vacuum system is shown in fig. 3.8.

3.6 Oven techniques

As we shall discuss in chapter 4, it is often convenient to mix the material to be matrix isolated with the matrix gas before deposition of the matrix. This clearly cannot be done with reactive species and other methods must be used. An important part is played here by ovens, and it is convenient to discuss the types of oven most often used here.

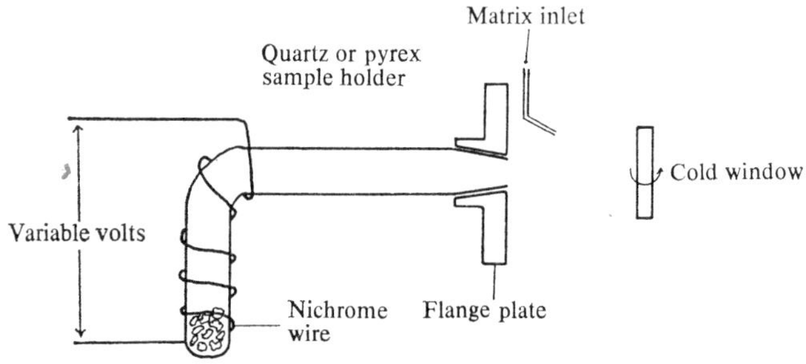

Fig. 3.9. Simple variable temperature furnace.

The basic technique used is that of co-condensation, in which the stream of matrix gas from the spray-on nozzle is joined by a stream of vapour effusing from an oven and the resultant mixture condensed on the cold plate immediately. A typical experimental arrangement is shown in fig. 3.9.

This can clearly be used for samples with the correct vapour pressure whatever the temperature, but we shall not discuss co-condensation of samples volatile at or below room temperature here, except to point out that the evaporation rate is best controlled by maintaining a steady temperature at which the vapour pressure is in the range 10^{-2}–10^{-4} mmHg. A large orifice can then be used.

The use of ovens begins when it becomes necessary to heat the sample to attain the necessary vapour pressure. For comparatively low temperatures a bath of hot liquid may surround the sample tube; heating may be external to the bath or through an immersed heater. Somewhat higher temperatures may be achieved using a resistance wire wound around the sample tube, with or without insulation. Temperatures up to about 1000 °C may be attained with nichrome wire, a quartz sample tube and plenty of insulation. It becomes difficult, however, to prevent the orifice from clogging with condensed material, as it is not heated directly. Above about 800 °C more elaborate ovens are needed.

Any method used for heating a sample must overcome a number of problems. These include:

(i) the sample container must not itself vaporise or react with the sample or any other material in contact with it;

(ii) the amount of radiation reaching the cold window from the hot parts of the oven must be minimised;

(iii) the transfer of heat from the oven itself to its surroundings by any method (radiation or conduction) must be minimised, not only to prevent damage to vacuum seals and the like but also to maintain the efficiency of the oven and maximise the sample temperature.

The sample itself must be carefully prepared for high temperature studies. In particular it must be free from more volatile impurities, since these would vaporise preferentially and might mask the spectrum of the sample or prevent the formation of the desired species. The sample container and all hot parts of the oven must be similarly clean; even so, traces of water, carbon dioxide and so on arise from fingerprints and similar minor contamination of surfaces during sample preparation and loading the oven. These, and carbon monoxide released during the heating of metal parts of the oven, can effectively be removed only by prolonged degassing under high vacuum at a temperature just below that required for sample evaporation before cooling down the apparatus and beginning deposition of the matrix.

Wherever possible the sample is arranged so that evaporated molecules have a straight-line path to the cold window. If this is not done, a considerable amount of material will condense at the obstruction, unless this is heated to a temperature above that of the oven.

The design of sample containers depends on the particular application. The material of which they are made limits the attainable temperature as the container must not melt, vaporise or react at the temperature used. Table 3.1 shows the melting points of some commonly used sample-container materials. Where the vapour pressure further limits the maximum operating temperature, this is also shown in the table. The apparent vapour pressures will depend on the purity of the material to some extent, and on the efficiency of degassing. Some materials, notably steels, are notoriously difficult to degas effectively. The reactivity of materials at high temperatures has been little investigated, and suitable materials for evaporation of a given sample are usually determined empirically.

There are two basic types of sample container, one useful for simple evaporation from a solid, in which the effusion orifice is large so that rapid evaporation is possible, and the other for cases where reaction between two solids or a solid and a vapour is necessary to produce the desired species (see fig. 3.10). In this type the orifice is small, so that the pressure in the sample container is kept not much below the equilibrium value and chemical equilibrium can be attained. The effusing gas is then a representative sample of the chemical equilibrium vapour of the system. Either type may be adapted for samples

TABLE 3.1. *Melting points and vapour pressure limits of some sample-holder materials*

	Melting point/K	*T*(vapour pressure limit)/K
W	3680	3000
Ta	3250	2850
Mo	2880	2400
Pt	2050	2000
Cu	1360	1300
C	4000	2500
Al_2O_3	2325	2000
SiO_2	2000	1600

that remain solid at all times, or for samples where melting takes place and a pool of liquid forms. Such sample cells are then heated in a furnace, which is usually heated electrically (either resistive or inductive heating may be used). For very high temperatures the sample may be heated directly, either by concentrating an intense light source on it or by bombarding it with a beam of electrons or high energy ions. These methods concentrate the heating effect on a very small spot, which reduces the problem of supporting and insulating a furnace.

3.7 Resistively heated furnaces

A simple resistance furnace is shown in fig. 3.10. The heating element consists of a coil of 0.5 mm tungsten or tantalum wire, supplied from a conventional transformer. The temperature is varied by altering the voltage supplied to the transformer; maximum power obtainable with normal mains supply and a variable transformer of 6 A capacity is 1500 W, supplied to the heating coil as 30 A at 50 V. The sample container is supported by the heating coil, which in turn is supported only by its electrical connectors. These are water-cooled copper tubes passing through an insulating plug in the vacuum jacket surrounding the furnace.

The heating coil is surrounded by concentric radiation shields, and the radiation emitted in the direction of the cold window is also trapped by shields, only a small hole in each leaving a direct path for evaporated sample. The radiation shields are made of refractory metals such as tungsten or tantalum, kept clean and highly reflecting, and should be supported from the outside wall of the furnace by

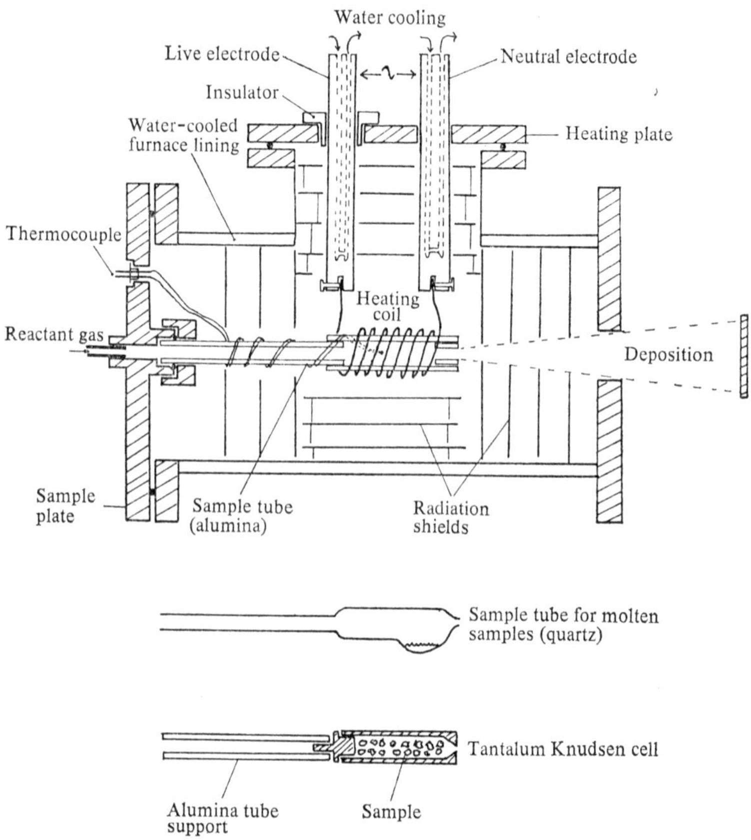

Fig. 3.10. Low current resistance furnace.

non-conducting material. Thus, the inside shields remain very hot, while the outer shields and the walls of the furnace assembly are cool. If necessary, the outer wall and the flanges carrying vacuum seals can be water cooled. Sample temperatures up to about 2500 K can be maintained with good shielding in this furnace; without shielding, the temperature attained is about 500 K lower.

These temperatures are well below the melting points of the coil materials (tungsten 3680 K, tantalum 3250 K), but cannot easily be exceeded, as failure of the coil occurs by over-heating at points of high resistance (for instance, kinks produced in winding, or other imperfections). The general temperature of the coil must then be kept well below the melting point to avoid failure, while the temperature

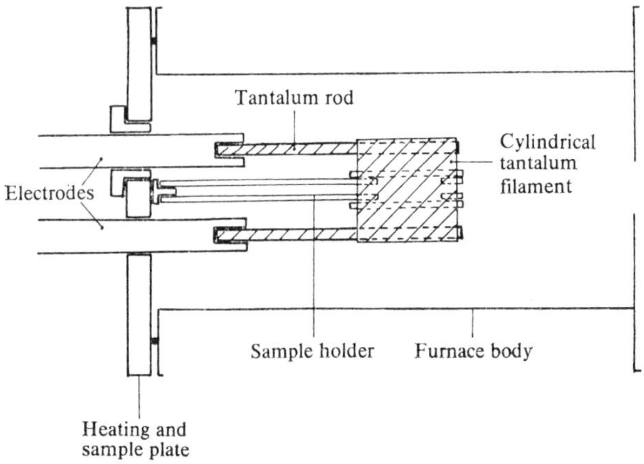

Fig. 3.11. High current tantalum resistance furnace.

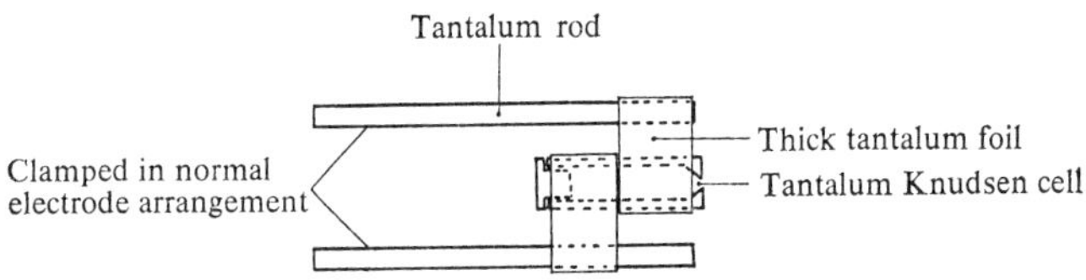

Fig. 3.12. In-circuit sample heating (high current).

of the sample container may be several hundred degrees lower than that of the coil.

A modified form of this furnace uses a cylinder of metal foil in place of the wire coil (see fig. 3.11). The likelihood of failure by overheating is thus lowered, while the temperature of the sample container more nearly approaches that of the heater because it is more effectively surrounded. A major disadvantage of this arrangement is that the foil has much lower resistance than the coil of wire, so very high currents are used. Typical conditions involve a supply of 400 A at 5 V, giving a maximum temperature of some 2700 K in the sample container. The power must be supplied to the heater through massive copper bars of about 1 cm^2 cross-section, and a much larger and more expensive transformer is needed.

A further modification uses a metal sample container as part of the heating circuit (see fig. 3.12). Very high currents are again required, and the power is delivered to the sample container through massive welded connectors. Temperatures of up to 2850 K are attainable in this way.

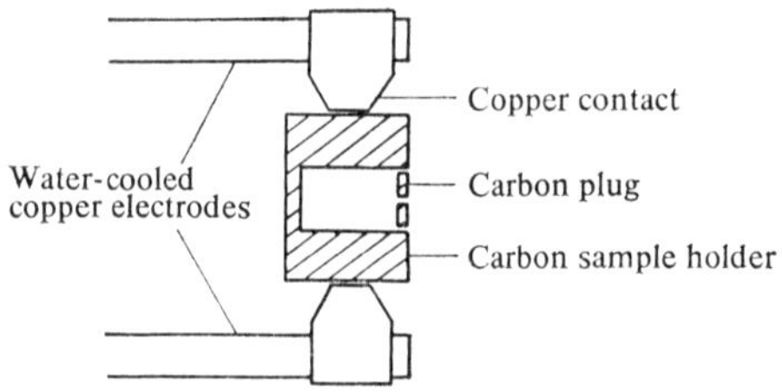

Fig. 3.13. Carbon resistance furnace (medium current).

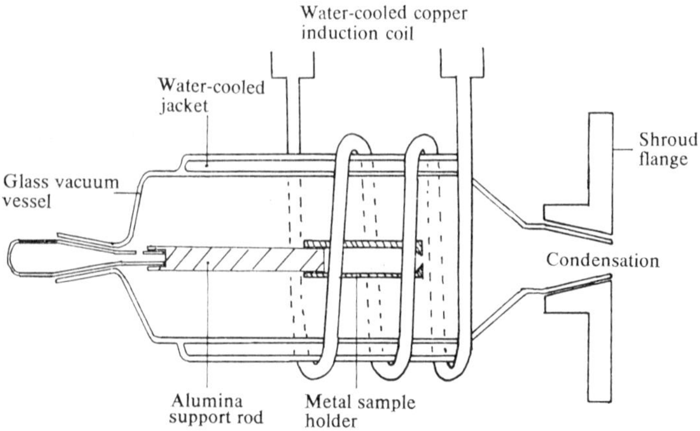

Fig. 3.14. Induction heating.

Lower current supplies can be used if the sample container is made of graphite (see fig. 3.13) whose resistance is greater. However, the vaporisation of carbon limits the temperature in practice to 2500 K.

3.8 Non-resistive heating

For the comparatively few studies involving evaporation temperatures above 2800 K, other means of heating have had to be used. The induction furnace (fig. 3.14) is capable of delivering large amounts of power without the need for direct electrical connections; the power coils may be cooled with water and remain cold. Heating of the metal sample container by induction can raise the temperature to well over 3000 K; radiation shields used in such a system must be non-metallic, otherwise they too would be heated, while the same applies to the vacuum jacket around the sample container.

Photo-heating. An even more subtle way of heating a sample is to focus an extremely intense beam of light, produced by a laser or a high power discharge lamp, directly on to it. This results in very local heating, and the main part of the sample remains comparatively cool, so the problems of supporting and insulating an oven are evaded. Reaction of heated sample with the sample holder is also prevented. Evaporation takes place from the surface of the sample and even small amounts of surface impurities can be studied in this way. On the other hand, unsuspected impurities can lead to unexpected results!

Bombardment heating. The sample can also be heated by bombardment with beams of high energy ions or electrons. Thus if the 'plasma' produced in a microwave discharge in the stream of matrix gas is allowed to flow over a sample, the excited atoms and ions produced in the discharge may lose their energy during a collision with the sample surface, ejecting small fragments in the process. Silica has been shown to lose SiO in this way, and this molecule has been detected in matrices deposited after such treatment. No reports of subsequent reactions of ejected species seem to have appeared, but it seems likely that such reactions may occur.

The microwave plasma is difficult to direct on to a surface, and only a very small proportion of the energy supplied to the discharge is used in evaporating material from the sample. Much more efficient bombardment is possible if an electron beam, which can readily be produced, accelerated, focused and directed, is used. In this case the heating effect can be extremely local, so problems with the sample container are minimised. It is possible to vaporise materials at temperatures exceeding 3000 K in this way. The high intensity electron beam may lead to the production of charged species among the products of evaporation, so the results obtained in this way are not directly comparable with those involving more indirect methods of heating.

Temperature measurement. There are only two practicable means of measuring the temperature of a hot sample isolated in a vacuum. For temperatures up to about 1800 K a thermocouple may be used, but above this temperature only optical pyrometry (in which the visible radiation emitted by the sample is compared with that from a resistively heated tungsten filament) can provide reliable results. The technique, involving visual comparison while the temperature of the filament is varied by changing the power supplied to it, is suprisingly accurate, perhaps because both the overall intensity and the effective 'colour' of the radiation emitted changes with temperature.

4 Production of matrices containing reactive species

We have considered what materials may be suitable for matrices to contain reactive species, and some of the technical details of apparatus needed for making such matrices. It is now appropriate to look in more detail at the methods available for making reactive species and how these methods may be adapted to matrix isolation. There is a clear primary distinction to be made between methods that generate the reactive species before deposition of the matrix (so that the reactive species itself is trapped during formation of the matrix) and methods that generate the reactive species within the matrix. We shall refer to the two types as involving *trapping* and *generation in situ* respectively. In addition, there are some instances of the use of a combination of methods of the two basic types.

We may take it as axiomatic that reactive species must be prepared from some more stable starting compound, or *precursor*. (The study of matrix-isolated stable molecules does not concern us here.) Energy must therefore be supplied to convert the precursor into the reactive species of interest. The most usual methods for supplying energy to the precursor prior to trapping involve *discharges* in flowing gases or evaporation from solids in *ovens*. *Chemical reactions* could in principle be used to generate reactive species in this context but seem not to have been much used. If the precursor is once isolated in the matrix it is quite well protected from any outside influence except radiation, and generation *in situ* of reactive species almost inevitably involves *photolysis* of the precursor. Subsequent *chemical reaction* is likely to occur only if diffusion of an atomic species is possible, or if the matrix cage contains reactive molecules.

4.1 Trapping methods

Oven evaporation. In some experiments the reactive species is generated by simply heating a solid in an oven, causing evaporation. The desired species then escapes through an orifice and travels on a straight-line path until it strikes the cold window, where it condenses.

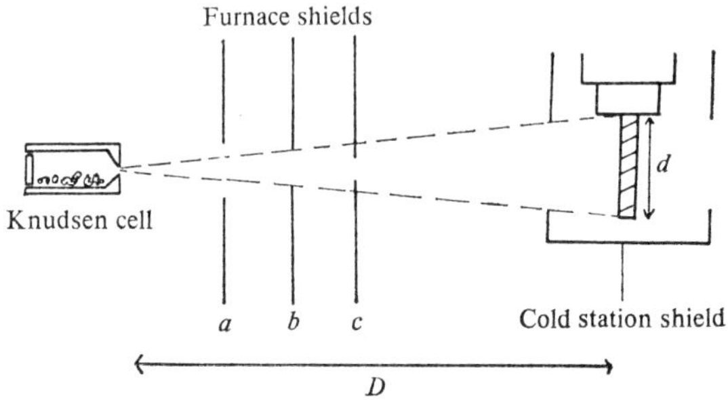

Fig. 4.1. Geometric factor and placing of radiation shields.

This condensation must occur at the same time as deposition of a much larger amount of matrix gas, and under conditions where the matrix rigidifies rapidly, if aggregation of the reactive species during deposition is to be avoided. In some cases it is convenient to pass the matrix gas through the oven so that the reactive species arrives at the cold window well mixed with matrix gas, but this greatly increases the heat flow required to condense the matrix, as all the matrix gas is now hot. It is more usual to co-condense the matrix gas with the vapour from the oven by arranging for the two beams to strike the window simultaneously. Co-condensation of matrix gas with two or more evaporated beams or co-condensation of a mixture of matrix gas and a stable precursor with an evaporated beam are simple extensions of this technique.

The *collection efficiency* for a vapour evaporated from an oven is related to the geometry of the system. The geometric factor is approximately $d^2/(d^2+4D^2)$, where d is the diameter of the collecting surface (the cold window) and D is the distance of this surface from the effusion cell. It is clearly most efficient to make D as small as possible, but this increases the problem of heat transfer to the window from the oven. As we mentioned in chapter 3, this may be reduced using metal shields with apertures for the evaporated beam. Careful design is needed if these are not to reduce the effective geometric factor: every relevant part of the collecting window should be accessible from the effusion orifice by a straight-line path. Fig. 4.1 shows correctly (*b*) and incorrectly placed radiation shields (*a* and *c*). The collection efficiency also involves a factor expressing the 'sticking probability', the proportion of molecules striking the cold surface that actually condense. For the matrix gas this can be as high as 0.6 or 0.8, and we may expect similar values for evaporated species

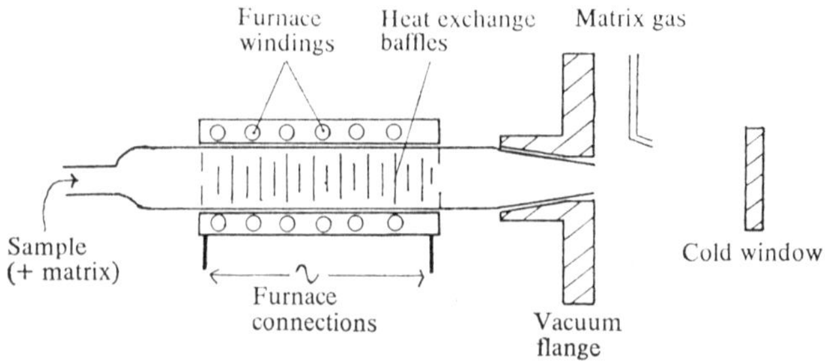

Fig. 4.2. Apparatus for pyrolysis.

in the presence of much larger amounts of matrix gas, despite their rather higher thermal energy. The geometric factor, with typical values of d and D (2 cm and 10 cm respectively), is much lower, of the order of 0.01. The material evaporated from the oven that does *not* reach the window should also be collected by the radiation shields and not allowed to increase the heat load on the cooling system by condensing on the heat-transfer block and other cold parts of the apparatus.

Problems that may arise in a co-condensation experiment include reaction or aggregation of reactive species before and during deposition. The lower the pressure in the beam of evaporated material, the lower the probability of bimolecular reactions before condensation becomes, and evaporation at the lowest practicable temperature is therefore advisable. The probability of reaction during deposition depends on the matrix ratio and the rate of cooling of the matrix material. Again, a low rate of evaporation is needed if these requirements are to be met simultaneously. If the vapour must be at a high pressure or super-heated to control the composition the effusion orifice may be made small so that slow effusion rates are maintained.

Pyrolysis. An even simpler experiment than that using an oven to generate reactive species is practicable when the desired species can be made by thermal decomposition of a volatile precursor that can be handled at room temperature. In this case a mixture of matrix gas and the precursor is prepared and passed through a heated tube before condensation of the matrix (see fig. 4.2). The presence of the large excess of matrix gas is usually helpful, as it reduces the mean free path of the unstable species and thus reduces the probability of wall reactions and aggregation before deposition.

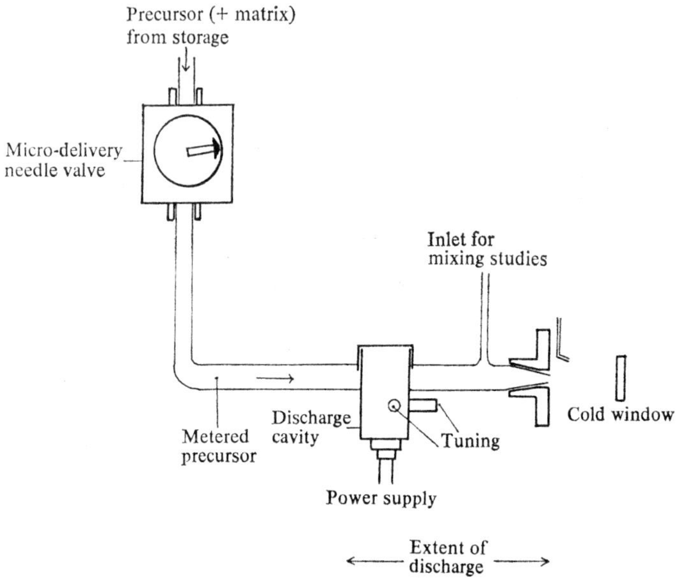

Fig. 4.3. Apparatus for discharge studies.

Discharge production. In these methods energy to split up a stable molecule is supplied electrically in such a way that a discharge, usually in the precursor itself, occurs. The most commonly used technique involves a microwave generator to supply power. The resulting discharge creates a high temperature plasma in a low pressure gas, and all but the most stable molecules seem to be completely dissociated to atoms. The reactive species of interest are then formed by recombination reactions outside the plasma before the matrix is deposited.

The experimental arrangement is shown in fig. 4.3. The discharge occurs in a glass or silica tube which passes through a resonant cavity connected to the microwave generator. It is necessary to cool glass discharge tubes with a cold air jet, as the heat of the discharge is partly generated in and partly transferred to the wall. Silica tubes can usually be used without forced cooling, because of the higher softening temperature. If reactive species formed in or near the discharge are to be collected in the matrix it is essential that they can reach the cold window without passing through constrictions in the tube, such as pressure-reducing valves. This means that the discharge must occur in a low pressure gas stream to avoid overloading the cooling system. It is then usually necessary to put a substantial amount of microwave power into the discharge simply to maintain the plasma. Typical experiments involve powers of the order of 100 watts. Most of this power must be lost before the gas stream reaches the cold window, or the matrix will not condense. The plasma loses energy as it flows down the tube outside the discharge, by radiation and heat transfer to the walls, and it is usually found that so long as the discharge zone is at least 10 cm or so away from the cold window the heating effect is not serious.

There are several basic situations possible in a microwave discharge experiment in connection with matrix isolation. The simplest experimentally involves mixing the precursor with matrix gas in the correct proportion and passing the mixed gas through the discharge and directly condensing the matrix. This is usually the preferred method. Molecular matrix gases, however, are not as stable as the rare gases, and would dissociate at least partially in the discharge. It is then necessary to carry out the discharge in the pure precursor, and to mix the discharge products with matrix gas before condensation. The experimental arrangement then becomes similar to that for a co-condensation experiment.

The third type of microwave discharge method involves discharging the matrix gas only, mixing the emerging gas with precursor (either alone or mixed with more matrix gas) and condensing the matrix after the mixed gas stream has travelled a few more centimetres. This is a much less destructive method of decomposing molecules, as energy is transferred from the discharge by excited atoms or molecules of the matrix gas. Each precursor molecule decomposed acquires the necessary energy in the course of a collision with one excited matrix atom, and the maximum amount of energy put into any one precursor molecule is thus limited to the amount carried from the discharge by one matrix atom. Argon is most commonly used in such experiments; argon atoms are found to have a metastable ^3P state about 1100 kJ mole^{-1} above the ground state. This metastable state can be populated in the discharge but cannot lose its energy by radiation, as the transition is forbidden.

Metastable argon atoms then survive for a considerable distance 'downstream' from the discharge zone, and can transfer their energy to precursor molecules that enter the stream of matrix gas. 1100 kJ mole^{-1} is more than sufficient to break any one chemical bond, but may not be enough to break all bonds in a precursor molecule of moderate size. In any case, it is likely that some or all of the fragments will carry off some of the excess energy as vibrational, rotational or translational excitation, or even electronic excitation in some cases. The most likely effect of collision of a precursor molecule with a metastable argon atom is the breaking of one or a few bonds to give a small number of fragments. These will, in general, be different from the products formed by atom recombination after the precursor has passed through a discharge.

The other rare gases also have metastable states that are populated in a microwave discharge; the amounts of excess energy and the lifetimes of these states decrease as the atoms become heavier. Nitrogen, another common matrix gas, also forms various excited species in the discharge, and some of them persist for a long time or distance 'downstream' of the discharge zone. While these could in principle be used to transfer energy to precursors in the same way, the situation is complicated by the formation of nitrogen atoms, which are chemically reactive and may be incorporated in the product fragments.

This fact is the basis of the last type of discharge method we shall mention, in which atoms formed by discharging nitrogen, oxygen or hydrogen are trapped in a matrix. Molecular nitrogen and oxygen are suitable matrix materials themselves, but hydrogen must be mixed with argon or some other matrix-forming material. Microwave discharges may be used, but it is found that d.c. discharges with metal electrodes in the gas stream give higher yields of free atoms. Such discharges using electrodes are not often used for decomposing less stable precursors to avoid possible complications with reactions between the electrodes and reactive products of the discharge. Fig. 4.4 illustrates the basic discharge experiments in matrix isolation studies.

Chemical reaction. Direct chemical reaction of two volatile precursors to produce unstable, reactive products is rarely used as a method of preparing species for matrix isolation, mainly because such reactions usually result in hot flames that are not easy to cool to matrix temperatures without destruction of reactive intermediates. It is usually necessary to produce a reactive *intermediate* by some means first, and to react this specifically with the precursor to give the desired product. The intermediate is usually formed by one of the methods outlined above, oven evaporation, pyrolysis or discharge. The precursor may be present while the intermediate is formed, or may have to be added afterwards if the conditions required to form the intermediate would cause decomposition of the precursor.

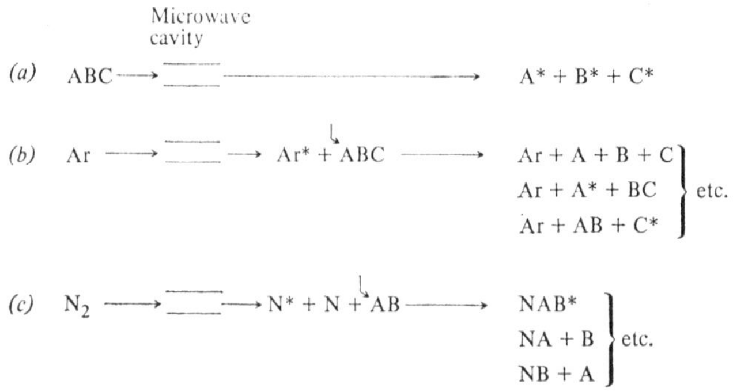

Fig. 4.4. Possible routes for production of species by microwave discharge.

Often it is possible to prepare the intermediate and to react it with the precursor in the presence of matrix gas, and the matrix can then be deposited directly. An example would be the formation of radicals, by thermal decomposition of bis(t-butyl)peroxide followed by abstraction of hydrogen from the precursor to give a radical product. If the intermediate is an atom it may be produced by a discharge reaction and the precursor introduced into the flowing gas stream outside the discharge zone. In this case the precursor would be decomposed if allowed to pass through the discharge.

Reaction of a volatile precursor with a heated solid is also used in some cases to produce reactive species; the matrix gas is usually mixed with the precursor before reaction. An example is the formation of SiF_2 by passing SiF_4 over elemental silicon heated to 1100 °C.

4.2 Generation *in situ*

In many ways it is easier to prepare a matrix containing the more-or-less stable precursor, and to decompose the precursor in the matrix. The problems of heat transfer from an oven to the matrix and of loss of reactive species by wall collisions and aggregation before and during deposition are thereby avoided, and it is much easier to ensure thorough mixing of matrix gas with a stable precursor than with a rapidly flowing gas stream containing reactive species.

This last consideration is important, as the importance of near-neighbour interactions is bound to be increased by any unevenness of distribution of matrix-isolated species. It is a common-place of elementary physics that a gas expands to fill the volume available to it, and it is usually tacitly assumed that this process is an instantaneous one and is independent of the presence of another gas. It is found, on the contrary, that the rate of the mixing process depends on the gases involved, and diffusion through heavy gases like xenon is particularly slow. Adequate mixing requires either a long time or a very turbulent gas flow if diffusion is slow, and it is more satisfactory to allow the matrix gas and stable precursor to mix in a large bulb than simply to inject one gas into a stream of the other. Uniformity is easier to achieve in a spherical bulb than in a cylindrical vessel.

Once the precursor and matrix gas are thoroughly mixed, the matrix may be simply deposited in the same way as a pure rare gas matrix with no added species. The normal precautions to prevent air and water from entering the matrix must be observed, but the stable nature of the precursor makes it unnecessary to protect the gas stream from contact with reactive glass or metal surfaces. The flow rate can then be controlled by a needle-valve, and the rate of deposition monitored by observing the fall in pressure on a mercury manometer or other device.

Deposition of a matrix can also be carried out using the pulsing technique described in chapter 2. This method is faster than the more conventional slow-flow technique, and can give matrices that are better annealed and optically clearer. Careful premixing of precursor and matrix gas is still essential if this method of deposition is used.

Once the precursor is isolated in the matrix it is quite difficult to decompose it by any means other than photolysis, unless the matrix also contains some reactive intermediate that can react chemically with the precursor after diffusion. Other methods involving bombarding the matrix with high energy particles are best thought of as special types of photolysis.

Ultraviolet photolysis of precursor in the matrix. The photolysis of a precursor in a matrix can only be efficiently carried out using radiation that is strongly absorbed by the molecule, and that has sufficient energy to break chemical bonds. In a few cases it may be possible to photolyse a coloured precursor using visible radiation, but in general ultraviolet radiation is required. A wide variety of ultraviolet lamps is available, and we shall discuss some of the most useful.

All ultraviolet lamps depend on an electrical discharge through a vapour to excite the ultraviolet radiation. The characteristics of the radiation emitted are controlled by the nature of the vapour and its pressure. Low pressure lamps tend to produce predominantly atomic emission lines, because the vapour is largely behaving as isolated atoms under the high temperature/low pressure conditions. Thus, if the lamp is filled with hydrogen at low pressure the emission is largely the atomic hydrogen line at 121.6 nm, while if it is filled with mercury vapour the emission lines at 184.9 nm and 253.7 nm are excited. High pressure lamps on the other hand give emission spectra consisting of broader bands, which may be continuous or may be made up of many lines. In either case, the effect is to give a broad spectrum of emission. The high pressure hydrogen lamp gives a many-line spectrum below 160 nm and above 500 nm, as well as a true continuum between 160 nm and 500 nm. The high pressure mercury lamp gives broad bands over the range 230 nm to 470 nm.

The windows through which the radiation passes between the discharge and the matrix (including the lamp envelope) are of great importance in governing the transmission of various wavelengths. Glass transmits only a little way into the ultraviolet and is generally useless for photolysis experiments. Sodium chloride and potassium bromide both transmit radiation down to 200 nm wavelength and make cheap windows for irradiation, but they fog easily in damp conditions and are too weak mechanically to be used for lamp envelopes.

Quartz is preferred for this purpose as its high strength and chemical and mechanical stability suits it for use in discharge tubes as well as for windows. The purest samples transmit down to about 160 nm. Below this wavelength only some crystalline fluorides, notably MgF_2 and LiF, form useful windows. LiF excels in this respect, the purest samples transmitting down to about 100 nm. Unfortunately, it is subject to fogging and must be repolished frequently to extremely high standards if it is to transmit efficiently. It is not sufficiently robust to withstand contact with a discharge plasma, and must therefore be used as a window on the end of a tube made of some other material.

It is often convenient to use a set of *filters* in conjunction with windows of more than adequate transmission so that bands of radiation of different energies can be tried in a given experiment without changing the windows involved in the vacuum sealing. Thus a moderately high pressure hydrogen lamp with LiF windows will provide a source of radiation between 100 nm and 1000 nm. The use of glass, KBr, quartz and MgF_2 filters enables the effects of various portions of this very wide spectrum of radiation to be investigated.

A complication inherent in the use of ultraviolet radiation of wavelength less than 200 nm is that atmospheric oxygen absorbs such radiation strongly. As the matrix is contained in an evacuated apparatus in any case it is comparatively simple to ensure that the radiation leaves the lamp window within the vacuum vessel and does not need to traverse any length of atmosphere. If longer-wavelength radiation is used it is more usual to leave an air gap between the lamp window and the window in the vacuum jacket of the matrix apparatus through which the photolysis radiation enters. This makes it easier to adjust lenses and filters in the light-path, and reduces the problem of heat transfer from the discharge to the cold window (by infrared emission from the lamp and its envelope). On the other hand it requires an extra window between the discharge and the matrix, as if the lamp is included in the vacuum sealing of the vessel the same window can do service as both lamp window and photolysis window.

The experimental arrangement for photolysis, then, is rather simple. An aperture in the various heat shields surrounding the matrix is provided, together with a window sealed to the outer vacuum case. The lamp is mounted so that radiation is directed on to the matrix through the window and aperture (see fig. 4.5). If it is possible for the photolysis and spectroscopic beams to be arranged at a rather small angle without interfering, one may monitor the disappearance of the precursor or the growth of the product without interrupting the photolysis (fig. 4.6). Often it is more suitable for the photolysis window to be at right angles to the spectroscopic windows, in which case the matrix and cooling system must be rotated for photolysis, and then back again before the progress of photolysis can be monitored spectroscopically.

For e.s.r. experiments, in which spectroscopic measurements must be made with the matrix in a cavity containing complex magnetic and microwave equipment, the matrix and its support are often *lifted* clear of the magnet region for photolysis and then lowered while the spectrum is monitored. These rotation and lifting motions are made possible by O-ring seals between concentric metal cylinders, which permit the cylinders to rotate in the plane of the ring or to slide perpendicularly to it without breaking the vacuum seal (see fig. 3.6). The O-rings are designed to fit tightly between the sealing surfaces, being somewhat stretched by the inner cylinder and compressed by the outer one. The O-rings are fixed in grooves in one of the metal surfaces, and grease is often used to ensure that sliding of the other metal surface can occur freely.

While the experimental arrangement for photolysis may be quite

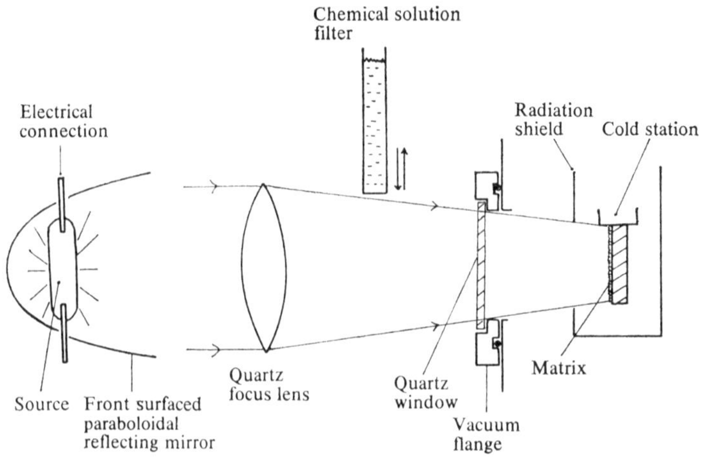

Fig. 4.5. Photolysis apparatus for matrix studies.

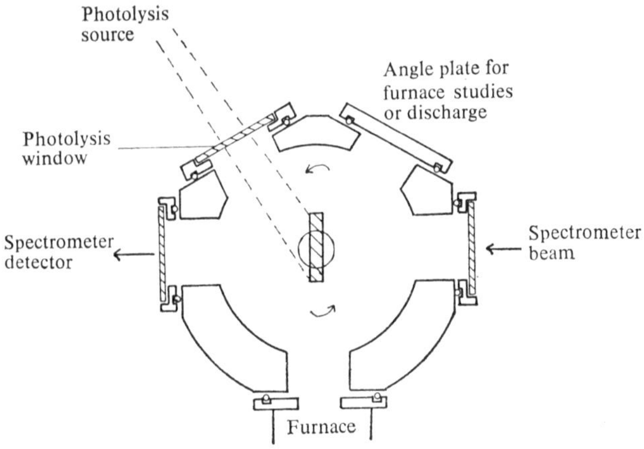

Fig. 4.6. Simultaneous photolytic monitoring apparatus.

simple, the processes involved are not, as an analysis of some of the possible results of photolysis at the molecular level will show.

Results of photolysis

In general, absorption is only probable if the photon energy of the incident light is just sufficient to raise the molecule to some electronically excited state. We shall call the ground state of the precursor *P* and the excited state reached after absorption of a single photon E. Such an absorption process corresponds to one of the electronic absorption bands of P; the probability of absorption will depend on the natures of the two states involved.

The excited electronic state E may behave in one or more different ways:

(*a*) It may *re-radiate*, emitting a photon of the same energy as the incident photon, and returning to the ground state P. This is probably the most likely result in all cases where the transition $E \longleftrightarrow P$ is allowed according to the selection rules for electronic excitation.

(*b*) It may *radiate a photon of lower energy* than the incident photon, so that the molecule ends up in a lower, but still excited state, L.

(*c*) It may *lose energy without radiating*, usually by transferring the excess to the matrix. Such energy must be conducted away through the matrix to the cooling system. Energy is usually accepted by the matrix in the form of 'lattice vibrations', low energy processes (of the order of 100 cm^{-1}) compared with electronic excitation (energies of the order of 50000 cm^{-1} or more). The transfer is then by no means easy, and considerable local heating may occur before the energy is effectively dissipated.

(*d*) It may *rearrange*, giving a new molecule I (an isomer of P) of different total energy from the initial precursor. Even if the isomer I has a higher energy than P it is most likely that some excess energy will have to be removed. This may occur by radiation or by transfer to the matrix. The latter process is the more likely, and the resultant local heating will usually allow the matrix to 'accommodate' to the changed shape of the molecule.

(*e*) It may *dissociate*, usually into two fragments, but occasionally into more. This may appear to be the desired end result, but two factors must be borne in mind: (i) the fragments are essentially *in contact*, being trapped in the site once occupied by the precursor; and (ii) one or both fragments are almost certain to be excited, that is to have a considerable amount of excess energy, as the initial photon energy is inevitably far higher than the energy required to dissociate the precursor.

The net result, then, is the formation of two more-or-less reactive species in close proximity with considerable excess energy. They will almost inevitably react together, reversing the dissociation, to give the initial excited state E, a lower excited state L, or an isomer I. If it is possible for energy to be removed rapidly from the system by

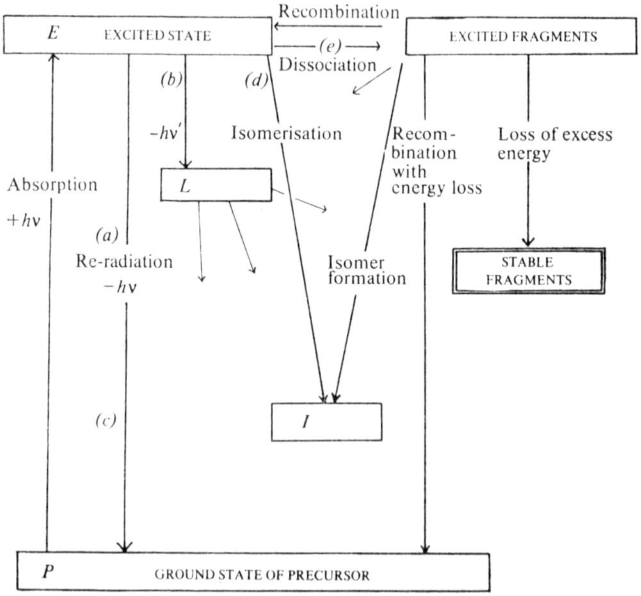

Fig. 4.7. Possible results of absorption of photon by matrix-isolated precursor *P*.

radiation or transfer to the matrix the ground-state precursor *P* may also be formed. Only if these processes can be prevented can an appreciable amount of fragments be stabilised. The various processes postulated here are set out in fig. 4.7.

Avoiding reversal of dissociation

There are several circumstances in which the reversal of dissociation may be avoided. Firstly, if one of the fragments is a *small atom*, which can diffuse readily in the matrix even without excess energy, it may escape from the matrix cage which contains the fragments before recombination can occur. Hydrogen atoms are ideal in this respect, and the photolysis of hydrides commonly gives high yields of fragments formed by loss of hydrogen. First row elements such as lithium, carbon, nitrogen, oxygen and fluorine may also escape by diffusion; in some cases it is necessary to maintain the temperature of the matrix within the annealing range to allow these heavier atoms to escape.

Secondly, if one of the fragments is an *inert molecule* it can behave after the dissociation as a new part of the matrix cage, and the only problem remaining is how to get rid of the excess energy. If this can be done, by radiation or by transferring it to the matrix, the remaining fragment will have a chance of persisting. An outstanding example of such an inert fragment is N_2, which, once formed, is not likely to be successfully attacked except by the most reactive of fragments.

As an example, we may consider the photolysis of an azide, RN_3. This appears to lead initially to the nitrene (RN:) by elimination of N_2. The nitrene may rearrange, especially if there are hydrogen atoms α to the nitrogen, giving an imine:

e.g.
$$CH_3N_3 \xrightarrow{h\nu} [CH_3N{:}] + N_2$$
$$[CH_3N{:}] \rightarrow CH_2{=}NH.$$

The product nitrene is not observed in this case, but where α-hydrogens are absent the rearrangement cannot occur and the nitrene is stabilised.

In some cases even nitrogen is too reactive to survive in close proximity to the other fragment. Thus photolysis of diazomethane, CH_2N_2, in an argon matrix does not produce the expected methylene, CH_2, in any detectable amounts. It is believed that CH_2+N_2 are indeed formed on photolysis, but that they immediately react to reform diazomethane. This is confirmed by experiments where ketene, CH_2CO, is photolysed in a nitrogen matrix. Again no methylene is observed, but the precursor is gradually converted to diazomethane during the photolysis. It seems that the initial photolysis of ketene to methylene and carbon monoxide is followed by reaction of the product methylene with matrix nitrogen, yielding diazomethane. Reversal to ketene will also occur, but N_2 by far outnumbers CO:

$$CH_2CO \overset{h\nu}{\rightleftharpoons} CH_2 + CO$$
$$CH_2 + N_2 \rightarrow CH_2N_2.$$

There are other, more puzzling, cases where dissociation occurs into two fragments which remain in reasonably close proximity, yet do not recombine. Two possible explanations have been put forward for such behaviour:

(i) That the fragments move apart after the dissociation step, so that the matrix around them is disrupted. As soon as all the kinetic energy of the fragments is exhausted the matrix, which will still be disordered and will have been unable to dissipate all the energy it has absorbed, will 'set' around the fragments in their final positions, so that they are trapped too far apart for recombination.

(ii) That the fragments formed initially are unlikely to be in their lowest possible electronic states. The fragments will therefore radiate energy and adopt their most stable possible states. It is now possible that the resulting ground-state fragments will be unable to recombine because they have either the wrong symmetry properties or too little energy. Thus the fragments are stabilised by electronic incompatibility rather than by distance.

Either explanation could be applied to some cases, such as that of the photolysis of transition metal hexacarbonyls, $M(CO)_6$. These are photolysed to $M(CO)_5 + CO$, which do not recombine. However, if the matrix is irradiated with low energy radiation (red or infrared light) the fragments disappear, being converted back into precursor. It can be argued that the CO initially formed has enough kinetic energy to 'burrow' into the matrix cage until it is trapped, too far from the $M(CO)_5$ fragment for recombination to occur, by the matrix resetting after being disturbed. Irradiation with low energy light then provides energy to disrupt the matrix so that the fragments can approach and recombine.

Against this explanation is the fact that the matrix itself, even when distorted by trapped species, does not absorb radiation that is effective in causing recombination. The energy must therefore be taken up by the $M(CO)_5$ fragment and transferred to the matrix indirectly.

The electronic incompatibility explanation points out that the initial photolysis is likely to lead to carbon monoxide in its ground state, $^1\Sigma$, but to $M(CO)_5$ in a state with no electrons in the sixth sigma orbital that was used in bonding the ejected CO in the precursor. This (a $^1A_1$ state) is most unlikely to be the ground state of the fragment, which would be expected to have a single electron in this orbital and a vacancy in one of the metal d-orbitals. The ground state would then be ^3E or $^3B_2$, depending on which d-orbital level was incompletely filled.

If the fragment, formed in an excited $^1A_1$ state, is able to lose energy and adopt the ground-state electron configuration before the recombination can occur it is unlikely that the two ground-state fragments could react. The $M(CO)_5$ fragment has lost energy, so that the system as a whole no longer has so much excess energy, while the symmetry of the fragments is such that reaction is not allowed. The effect of the low energy radiation is now to excite the ground-state $M(CO)_5$ to its $^1A_1$ excited state, which can react with CO to reform $M(CO)_6$. It would seem that a simple test of these alternatives would be to raise the temperature of the matrix until annealing took place, when fragments trapped too far apart to recombine might be able to approach each other.

The effects of continued photolysis

Returning to our analysis of the results of photolysis (see fig. 4.7), we may produce new species from our precursor *P* if either process (*d*) or process (*e*) is a reasonably probable one, and if the products are not reconverted spontaneously to the precursor. Processes (*a*) and (*c*) lead back directly to the precursor, which is then available for a further excitation, while process (*b*) is essentially equivalent to the result of photolysis with a lower energy photon and may have results similar to (*a*), (*c*), (*d*) or (*e*). Prolonged irradiation should apparently result in complete conversion of the precursor to products, even if most individual precursor molecules have to be excited many times before forming the desired product.

Several factors prevent this ideal situation from being attained. Firstly, the desired fragmentation may be but one of several possible modes of decomposition of the excited state *E*, so that by-products are formed in some definite proportion of cases. It may be possible to influence this proportion by changing the energy of the exciting photon. Often one may only be able to achieve the conversion of a few per cent of the precursor to the desired product because of competing reactions leading to by-products.

The second factor we must consider is the possibility of absorption of the exciting radiation by the desired product leading to its decomposition. If both the precursor and the product absorb the incident radiation, the rate of increase of concentration of product depends on the difference between the rates of production and decomposition. If the probability of decomposition equals or exceeds the probability of formation no build-up of concentration can occur. In any case, only a proportion of the precursor can be converted to the desired product; in some circumstances a 'steady-state' concentration can be built up that will remain constant while the rate of decomposition equals the rate of formation of product. It is sometimes possible to find a wavelength of light absorbed by the precursor and not the product, which avoids the difficulty.

Further photolysis of an initial product can be turned to good account if the second product is also a species of interest. An example occurs in the photolysis of methyl azide mentioned above; photolysis of the initially observed product, $CH_2{=}NH$, leads to loss of hydrogen and formation of HNC, an isomer of hydrogen cyanide. In some cases the second step is achieved after prolonged irradiation at the original photon energy whereas in other cases irradiation at a different wavelength is required for the second step. Occasionally the effect of the second irradiation is to reverse the effect of the initial decomposition, or to transform an unstable isomer into the more stable form.

Other types of exciting radiation. These are rarely used, mainly because ultraviolet radiation is so easy to produce and use. *X-rays* can be produced quite readily and have been used to irradiate matrix-isolated samples; extensive decomposition was certainly caused and some fragments were formed in electronically excited states. Ions were also detected among the matrix-isolated products. γ-rays have been very little used for optical studies, but are often used to create fragments for e.s.r. studies. They are formed by nuclear decay reactions, so are not readily controllable.

Charged particles (electrons, protons, heavy ions) are usually found to be rather too energetic for matrix studies, as the cooling system has to cope with the heat released from the particle beam as it is stopped. Again, these are rather indiscriminate as energy sources, and lead to extensive fragmentation, excitation and ionisation. A few examples of their use will be given in chapters 7 and 8.

4.3 Generation *in situ* by chemical reaction

Reaction of precursor with mobile reactive species. As we have discussed in chapter 2, atoms and small molecules are often found to become mobile in a matrix at a temperature below that at which wholesale diffusion of large molecules is possible, so the reaction of a static large precursor with mobile small species can occur under conditions where the resulting reactive product is unable to diffuse. The reaction of an atom with a molecule can in principle be of one of at least three types, which may be described as:

(i) Attachment $AB + C \rightarrow ABC$
(ii) Insertion $AB + C \rightarrow ACB$
(iii) Abstraction $AB + C \rightarrow A + BC$.

All three types have been observed in matrices; A and B here can represent either single atoms or larger groups.

The diffusing atoms may be introduced as the matrix is formed, by co-condensation, or by photolysis of a second trapped precursor. The co-condensation method is used mainly for alkali metal atoms, which can conveniently be evaporated from ovens. Specific precursors that are readily photolysed to give other atoms have been found empirically. Some notable examples are:

$$HI \xrightarrow{h\nu} H\ (\text{mobile}) + I\ (\text{static})$$

$$F_2 \xrightarrow{h\nu} 2F\ (\text{mobile})$$

$$N_3CN \xrightarrow{h\nu} C\ (\text{mobile}) + 2N_2\ (\text{static}).$$

Diffusion of the alkali metal atoms is readily controlled by changing the temperature of the matrix. If the spectrum is run as soon as the matrix is deposited, and while it is too cold for diffusion even of atoms, any changes that occur subsequently as the temperature is raised can be ascribed to reactions following annealing and motion of atoms. It is usually found that photolytically produced atoms have sufficient kinetic energy to begin diffusion immediately, whatever the overall temperature of the matrix.

The alkali metal atoms have been most extensively studied as reactive intermediates in matrices. They undergo, for example, atom-attachment reactions with matrix-isolated O_2 molecules on diffusion, to give species formulated as MO_2, M_2O_2 and MO_4. The metal atoms seem to become bonded symmetrically to both oxygen atoms of the O_2 molecule. With N_2O an atom-abstraction reaction occurs, giving MO and nitrogen. Further attachment of M to MO gives M_2O.

It was hoped that similar atom-abstraction reactions would occur when diffusing alkali metal atoms reacted with halides, and indeed the preparation of methyl radical by reaction

$$CH_3X + M \rightarrow CH_3 + MX$$

was reported in the literature. Unfortunately, further investigation showed that the reaction was better described as an atom insertion,

$$CH_3X + M \rightarrow CH_3MX$$

as the vibration frequencies characteristic of the methyl group were different for different alkali metals and for different halides. It seems that the atom-abstraction reaction fails if neither product is sufficiently mobile to escape from the matrix cage and neither is sufficiently stable to resist attack and chemical binding to the other.

It is quite possible for small molecules to diffuse in matrices at quite low temperatures, and they are known to undergo aggregation reactions under such conditions. No clear-cut example of the reaction of a diffusing molecule with a stable precursor seems to be known, however. Successive reactions of diffusing monomer with small oligomers are well established, and serve to illustrate the principle of the reaction of diffusing molecules. A well-studied example is that of SiO, which forms successively Si_2O_2 and Si_3O_3, until no free SiO remains. The rate of polymerisation then falls dramatically, as the larger molecules are effectively immobile. Other diatomic molecules with large dipole moments tend to behave similarly, while aggregation as such is the common fate of most matrix-isolated molecules eventually.

Reaction of matrix material with reactive intermediate. It is usual in matrix isolation to use a matrix material that is chemically inert, so that reactive species may be preserved unchanged. It is, however, quite possible to use a chemical reaction involving the matrix material to prepare a species for study, so long as the matrix does not react further with the product. It is then necessary to employ as an intermediate some species able to react with the matrix material.

The heavy rare gases, krypton and xenon, will react under some conditions with fluorine, giving difluorides, but the rare gases are otherwise completely inert. The diatomic substances most commonly used for matrices, nitrogen, oxygen and carbon monoxide, are very much more reactive, and a large number of reactions of these matrix materials with reactive species have been reported. In addition, it is now suspected that unassigned bands in many early studies may be due to products of reaction of nitrogen, oxygen or carbon monoxide, present as impurities in rare gas matrices, with reactive species. It is equally possible that bands due to such products have been erroneously assigned to other species. It has been estimated that residual nitrogen pressures below about 10^{-6} mmHg must be maintained during deposition of a matrix if the proportion of nitrogen in the matrix is to be kept below 1 %.

Free atoms are, of course, among the most reactive intermediates used in matrix studies, and their reactions with 'reactive matrices' have received much attention recently. Metal carbonyl complexes, the more novel metal dinitrogen complexes and metal dioxygen complexes have all been produced in this way; some examples are discussed in chapter 8.

Reactive matrices can also react with photolytically produced reactive intermediates. A classic example is that of methylene (see §4.2), which reacts with carbon monoxide or nitrogen matrices to form ketene or diazomethane respectively. Fluorine atoms react with carbon monoxide or oxygen matrices to give FCO or FOO respectively, but do not appear to react with nitrogen. Hydrogen atoms are also captured efficiently in carbon monoxide or oxygen matrices, giving HCO or HOO, but do not react with nitrogen.

Reactions of photoelectrons. It has recently been appreciated that high energy radiation, including some ultraviolet light, can cause photo-ionisation, that is ejection of electrons from atoms or molecules, in matrices. The resulting formation of a positive ion is no more than another example of photolysis, but the fate of the photoelectron is more interesting.

It is ejected into a hostile environment, as most matrix materials have very little affinity for electrons. (Oxygen is an exception; it has a real electron affinity, and will immediately capture any photoelectrons.) Some reactive species, on the other hand, do have an affinity for electrons, that is they form relatively stable negative ions. If such species are present when photoelectrons are produced they will capture them, and the matrix will ultimately contain both positive and negative ions.

Most studies using ultraviolet radiation are limited to an energy of about 6 electronvolts (eV) (about 50000 cm^{-1} or 600 kJ $mole^{-1}$), and very few substances have ionisation potentials as low as this. The formation of ions under such conditions is then most unlikely. Studies involving the hydrogen atomic emission line at 121.6 nm (H Lyman α), on the other hand, quite often produce ions, as the energy available is now about 10 eV. This is enough to ionise many stable molecules, and most reactive species (which tend to have lower ionisation potentials than stable molecules because they contain electrons in weakly bound orbitals).

If a copious supply of photoelectrons is required for a particular study it is convenient to trap alkyl amines or alkali metal atoms, both of which have low ionisation potentials, in the matrix together with the precursor. Several small polyatomic negative ions (e.g. SO_2^-, NO_2^-, O_3^-) have been produced in matrices in this way.

It has been shown that if two species, one a potential source of a photoelectron and the other able to capture an electron, are trapped close enough together in the matrix, their potential wells may overlap (see fig. 4.8) so that it is no longer necessary to supply the full ionisation potential to raise the electron over the potential barrier between the wells. The result of this process is then better thought of as an ion-pair than as two independent ions, while the process itself is a charge-transfer transition rather than ionisation followed by capture of the photoelectron. It is claimed that matrices containing

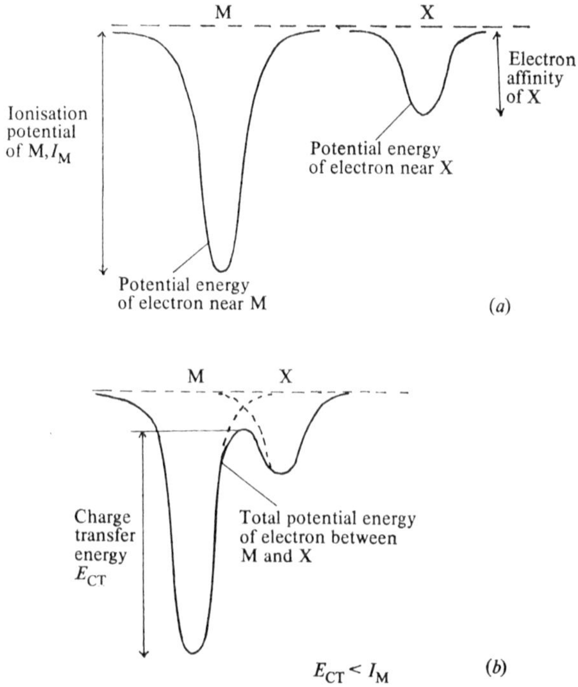

Fig. 4.8. Potential wells of two trapped species: (*a*) species far apart; (*b*) species close together.

both sodium atoms and polynuclear aromatic hydrocarbons show absorption bands corresponding to such charge-transfer processes at energies considerably lower than the ionisation potential of sodium. It is not clear how important such effects are in other systems, or how reasonable it is to consider ions produced in matrices as isolated from each other. It is clear, however, that systems that contain ions in matrices are almost always unstable relative to the corresponding systems without charge, and it is often found that ions that can be generated using very energetic ultraviolet light can be induced to revert to the uncharged species using much less energetic radiation.

As we mentioned earlier, the use of very energetic X-rays, γ-rays or charged particles to irradiate the sample often gives rise to ions. In some cases ionic fragments seem to be produced directly, while in others they arise from photo-ionisation reactions of neutral fragments.

4.4 Organic matrix isolation in glassy solids

The study of reactive organic intermediates (such as alkyl, aryl, acyl, azyl and oxyl radicals) by e.s.r. is one of the main sources of information about their structures. In such cases the almost inevitable presence of hydrogen atoms in a variety of chemical environments near the radical centre gives rise to splitting of the e.s.r. signal, and considerable information about the structures and bonding can be deduced.

It is possible for such studies to use less rigorously non-reactive conditions than those characteristic of inorganic matrix studies; the temperature of liquid nitrogen is often adequate, and rigid glassy matrices of organic substances that do not give radicals under the conditions used are suitable.

The use of solutions of precursors, which are then simply frozen to glassy solids, makes preparation of matrices extremely simple, and allows variation and measurement of concentrations to be carried out precisely. Typical solvents, chosen because they form glasses rather than crystalline solids when cooled rapidly to 77 K, are branched alkanes (e.g. 2-methylpentane), which are ideally inert. More reactive mixtures, which are better solvents for precursors, can be used in some cases. A favourite mixture is EPA (ether, pentane, alcohol), which consists of diethyl ether, isopentane and ethyl alcohol in the proportions 5:5:1.

Formation of radicals is achieved, usually after freezing, by irradiation with ultraviolet light, X-rays, γ-rays or electrons. In some cases radical initiators, such as di-t-butyl peroxide, are included to provide a reliable source of radicals on photolysis. Subsequent reactions with precursors (often accelerated by controlled thawing of the matrix) lead to products by atom transfer reactions and consequent rearrangements.

5 Application of spectroscopy to matrix-isolated species

We have now considered some of the technical aspects of matrix formation and the ways in which matrices containing reactive species may be prepared. It is pertinent at this stage to deal with the ways in which these reactive species may be characterised and studied.

A matrix-isolated species is indeed isolated, in the sense that we can only operate on it under conditions where the matrix is not disrupted. This essentially limits our means of characterisation and study to non-destructive spectroscopic methods. These may be further limited by the matrix itself, which must not interfere with the spectrum of the species under investigation.

5.1 Spectroscopic methods

The essence of most spectroscopic techniques is the absorption or emission of electromagnetic radiation in resonance with a transition between two discrete states of a molecule or atom. The two states may, for example, be different in the electronic, the vibrational or the rotational parts of the molecular wavefunction; for atoms only the electronic wavefunction is relevant. The energies of transitions involving electronic, vibrational and rotational changes respectively usually fall within the ranges shown in fig. 5.1, which also includes the corresponding regions of the electromagnetic spectrum.

Other transitions, which may be observed only in the presence of a magnetic field, are those involving the magnetic moments associated with the spin quantum numbers of nuclei or unpaired electrons. In the magnetic fields that can conveniently be generated in the laboratory, transitions of this type are observed using radiofrequency or microwave radiation, of less than 1 cm^{-1} energy. They give rise respectively to nuclear magnetic resonance (n.m.r.) or electron spin resonance (e.s.r.) spectra.

Nuclear quadrupole resonance (n.q.r.) is another spectroscopic technique using radiofrequency radiation of very low energy. It arises

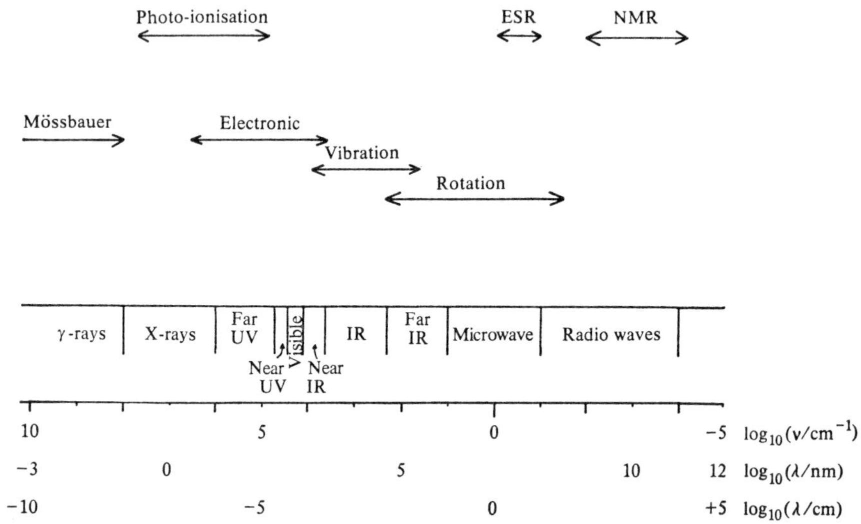

Fig. 5.1. Spectral ranges.

from the interaction of any quadrupole moment of an atomic nucleus with the electric field generated by the surrounding atoms, so needs no applied magnetic field.

At the other extreme of the electromagnetic spectrum, Mössbauer spectroscopy is a resonant process involving absorption of γ-rays, which have energies of the order of 10^8 kJ mole^{-1} (1 MeV), by atomic nuclei. The exact transition energy depends on the chemical environment of the nucleus, so the technique can provide information on chemical bonding as well as the presence or absence of a particular type of nucleus.

A few non-resonant spectroscopic methods may be used, though they usually give much weaker spectra than resonant methods. The most notable of these is Raman spectroscopy, where a photon of visible light that falls on the sample may be re-emitted as a photon of lower energy. The energy lost by the photon usually corresponds to a vibrational excitation of the sample, though rotational and electronic excitation by the Raman effect are also possible. It is also possible for the photon to emerge with a higher energy, if the sample is initially in a vibrationally excited state. This is rarely the case with matrix-isolated species.

The other non-resonant processes that are increasingly being used spectroscopically involve photo-ionisation, the ejection of an electron from the sample by a photon of more than sufficient energy. The spectrum of excess electron kinetic energies found can be related to the ionisation energies of the electrons in the sample. Two main types of this photoelectron spectroscopy have developed, one using X-rays with over 1000 eV (about 10^7 cm^{-1}) energy as ionising photons, the other using ultraviolet photons of 10–40 eV ($8–32 \times 10^4$ cm^{-1}) energy. X-rays are able to eject electrons from inner orbitals of the atoms concerned, and hence give information about the types of atom present, whereas ultraviolet excitation ejects electrons only from the valence-shell molecular orbitals, which are characteristic of the molecule present rather than of its constituent atoms.

Most of these methods suffer from severe disadvantages when applied to matrix-isolated species. Thus, such species can rarely rotate freely, so rotational spectroscopy as such is of little use. Nuclear magnetic resonance gives rise to broad signals of very low intensity if the sample is not fluid, because of the long relaxation time under such conditions. Nuclear quadrupole resonance spectra are also very weak and difficult to detect even with large samples. Neither technique seems to have been applied successfully to a matrix-isolated species.

The great disadvantage of Mössbauer spectroscopy is that only a few elements, notably iodine, iron and tin, give useful spectra, though other elements could in principle be used. A second disadvantage is that signals from different species usually overlap, as they occur over a very small range of energies. It is certainly possible in favourable cases to distinguish the presence of two species in similar concentrations but it seems unlikely that this technique will ever be able to give information about minor components of complex mixtures of species.

The non-resonant methods suffer very severe disadvantages because of the relative weakness of the signals, because of the difficulty of depositing large amounts of matrix-isolated material and the inevitable low concentration. Recently, high powered laser sources have made it possible to study matrix-isolated species by Raman spectroscopy and the technique is becoming a useful companion to infrared spectroscopy for the study of vibrations of matrix-isolated species. Photoelectron techniques have the additional disadvantage that the emitted electron is likely to be affected by the matrix, losing kinetic energy before it escapes. Indeed it is likely that only the surface layers of a matrix will be accessible to this type of study. The X-ray excitation method seems most likely to give useful results, as the higher energy of the emitted electron makes it less easy to capture.

By far the most useful types of spectroscopy for the characterisation and study of matrix-isolated species are electronic, vibrational and electron spin resonance. Electronic spectra, in the visible and ultraviolet regions of the electromagnetic spectrum, are studied in absorption or emission, vibrational spectra by infrared absorption or using the Raman effect. Electron spin resonance spectra are observed with microwave radiation in a powerful applied magnetic field. Adequate spectrometers of all these types are commercially available, and can be more-or-less readily adapted for use with matrix-isolated samples.

Closely related to the distinct physical phenomena that give rise to the types of spectroscopy listed above, there are distinct advantages and disadvantages related to our purposes of (*a*) characterising and (*b*) studying the structures and other properties of matrix-isolated species. These effectively limit the use of each type of spectroscopy to particular types of species. We shall discuss these limitations in detail in this chapter, and indicate where each type of spectroscopy can be particularly useful.

5.2 Electronic spectroscopy

Electronic transitions can, in principle, occur in any part of the electromagnetic spectrum, but most commercial UV/visible spectrometers cover a range of about 190–850 nm wavelength, corresponding to energies from 52000 to 12000 cm^{-1}. Below 190 nm atmospheric absorption makes it necessary to use evacuated spectrometers, and few matrix studies in this vacuum UV region have been reported. Very few electronic transitions of less than 12000 cm^{-1} energy are known, though reactive species with unpaired electrons and incompletely filled orbitals are more likely to give rise to low energy transitions than more normal molecules. The near-infrared region, between 12000 and 3000 cm^{-1}, might well prove interesting for matrix studies.

Necessary conditions for electronic spectroscopy matrix studies. If we are to use electronic spectroscopy to characterise a matrix-isolated species certain conditions must be satisfied:

(i) the species of interest must give rise to an absorption or emission spectrum in the region covered by the spectrometer;

(ii) the matrix material, the inner cold window and the outer windows must not mask this spectrum; and

(iii) the spectrum observed must be identifiable as due to a particular species.

The first condition usually limits observation to those species giving spectra between 190 and 850 nm; most reactive species will give bands in this region, though some very stable materials have no absorption in this range. This is fortunate, as it enables the second condition to be met without too much difficulty. Windows of quartz or salt (NaCl or KBr) are adequate in terms of transparency and mechanical strength. The common matrix materials are also transparent over the entire range (see chapter 2).

The third condition is more restricting, and indeed it effectively limits the use of electronic spectroscopy to those species whose spectra have been identified in gas-phase studies. This is because the only information directly related to the structure of the species responsible for an electronic transition is contained in the rotational fine-structure of the band, which can be analysed to give the moments of inertia. Other information, such as the band position, the intensity and the vibration pattern, cannot be related directly to the structure. Only if the spectrum is observed in the gas phase, where the rotation fine-structure is resolvable, and then compared with the spectrum found in the matrix, where only the band position and intensity and the vibration structure can be measured, may an identification of the species responsible for the matrix spectrum be made.

It is not by any means easy to make the identification certainly even then, as shifts in band positions and changes in vibration frequencies between gas and matrix spectra may be large. It is not safe to identify a predominant species from its infrared spectrum in a matrix and assign the electronic spectrum to the same species, as electronic spectra vary very widely in intensity, and the strongest electronic spectrum may arise from a very minor constituent of a mixture. In some cases minor impurities quite unrelated to the species expected may give strong electronic spectra, while the species of interest give no bands at all.

Spectral patterns. Electronic spectroscopy, though experimentally quite easy, is then not very useful as a means of identifying matrix-isolated species. If a spectrum can be identified more or less definitely as arising from a particular species of interest, it is often possible to obtain some information about vibration frequencies. We shall discuss the effect of the matrix upon vibration frequencies and electronic transition energies in chapter 6, but must point out here that the spectrum observed in a matrix may well *appear* quite unlike that found in gas-phase studies. This is because gas-phase spectra usually show bands arising from absorption by vibrationally excited molecules, since reactive species are generated at high effective temperatures. Matrix-isolated species, on the other hand, are trapped at very low temperatures and all absorption spectra arise from the lowest vibrational level in the ground electronic state. Fig. 5.2 shows how these differences affect the spectra observed.

The matrix *absorption* spectrum will then be expected to give clear information concerning an upper-state vibration frequency. For a species with more than two atoms there are several vibrations, but it is rarely possible to obtain the values of more than a few of these from the electronic spectrum. An *emission* spectrum, which may be excited by irradiating the sample with an intense light beam, may contain a progression of bands in a vibration frequency of the ground-state molecule. Few emission studies of reactive species in matrices have been reported.

Some species give only broad absorption bands, which are almost completely useless for identification purposes and give very little information about the species responsible. This is particularly common among larger molecules, but diatomics can give similar bands.

Scattering. One further practical problem is important in relation to the use of electronic spectroscopy in the study of matrix-isolated species. This is the phenomenon of scattering of radiation by the matrix. The difference between a clear and a cloudy matrix is at once apparent to the eye, and also to a spectrometer working with visible or ultraviolet light, but the exact conditions of deposition needed to give a clear matrix must usually be found empirically. Traditionally, slower deposition has been held to give clearer matrices, but the newer pulsed deposition technique can also give very clear matrices. The scattering is related to the sizes of the distinct crystals in the matrix; light whose wavelength is less than the grain-size is badly scattered.

5.3 Vibrational spectroscopy in the infrared

Vibrational spectroscopy too suffers from definite limitations in its usefulness for the identification of matrix-isolated species, but these are rather different from those attendant on electronic spectroscopy.

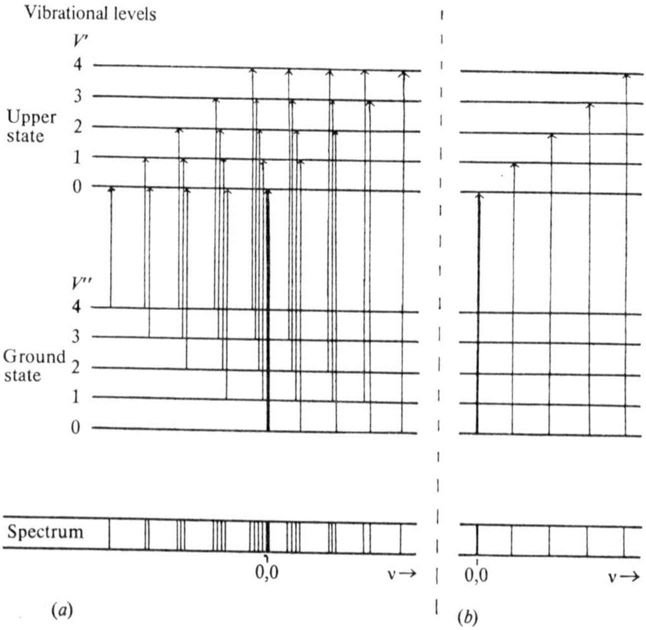

Fig. 5.2. Electronic spectrum of species in gas and matrix. (*a*) Gas: Boltzmann distribution over vibrational levels in the ground state. (*b*) Matrix: only zero level populated in the ground state.

In the first place, only two classes of substances (isolated atoms and homonuclear diatomic molecules) have no vibrational spectrum in 'the infrared'. Commercial spectrometers are available covering the range containing all known stretching vibrations (~ 4000–100 cm^{-1}); most deformation vibrations fall in the same range, though they are lower in energy than the stretching vibrations involving the same atoms. In practice the region in which KBr is transparent and can be used as a window material for the matrix and within the spectrometer ($\nu > 400$ cm^{-1}) has been far more extensively studied than the 'far infrared' below 400 cm^{-1}, but even the latter region is strictly available for study and is becoming more thoroughly investigated.

The ideal matrix material is rather more limited for infrared than for ultraviolet studies, only nitrogen, oxygen and the rare gases being completely transparent over the normal range. Even these materials have lattice vibrations in the far infrared (see chapter 2). In some cases it is possible to work with carbon monoxide, carbon dioxide or methane, or even carbon tetrachloride or sulphur hexafluoride, taking advantage of the clear 'windows' in their spectra to observe the weak bands of matrix-isolated species.

Window materials pose little problem in the normal infrared. Potassium bromide, which is cheap and readily available, is used down to 400 cm^{-1}, while caesium bromide or iodide may be used to lower frequencies (down to 250 or 200 cm^{-1}). The caesium salts are much more expensive, but mechanically rather stronger than potassium bromide. Interestingly caesium iodide transmits radiation down to 120 cm^{-1} when cooled to matrix temperatures. Below about 400 cm^{-1} polythene can be used, though it suffers from several disadvantages. It has an absorption band at 72 cm^{-1}, conducts heat very poorly, and is mechanically weak. Silicon and germanium windows may be used below 200 cm^{-1}, and quartz is also useful, transmitting far-infrared radiation below about 200 cm^{-1}. Most of these 'far-infrared' windows suffer the great disadvantage that they do not transmit in the normal infrared, so that comparison of the two regions must be made using two separate samples. For this reason potassium bromide or caesium iodide windows are preferred wherever possible.

One can be fairly sure then of obtaining at least one infrared band from any matrix-isolated species other than an atom or a homonuclear diatomic. The problems associated with the use of infrared studies arise in the assignment of individual peaks to particular species. Very few reactive species have been observed by infrared spectroscopy in the gas phase, so no direct comparison of frequencies can be made. In some cases the electronic spectrum of the species observed in the gas phase has been analysed to give the ground-state vibration frequencies, and these can be used to identify the peaks in the matrix infrared spectrum. This is possible only for a few diatomic and triatomic molecules that have been conclusively identified by their gas-phase electronic spectra. In all other cases bands must be assigned to reactive species trapped or produced in matrices more or less empirically using the methods described below.

Group frequency method. The first, least rigorous method, and the one most liable to error, is that using 'group frequencies'. This involves the use of certain general correlations that have been established, chiefly for stable molecules, between the groups present in a molecule and the infrared bands observed in its spectrum. Thus one may expect to find C–H stretching vibrations in the region 3300–2700 cm^{-1}, C–F stretches near 1200 cm^{-1}, P–Cl stretches near 500 cm^{-1}, and so on.

Less reliable correlations also apply to deformation vibrations. Unfortunately, many reactive species, especially small molecules, have vibration frequencies that fall outside the normal ranges, either because of their unusual electronic structures (which may weaken or strengthen bonds) or because of coupling between vibrational modes. Thus while the normal range for the Si–H stretching vibration is 2350–2050 cm^{-1}, the vibration frequency for SiH itself in a matrix occurs at 1967 cm^{-1}.

A further disadvantage of this method is that, in its simplest form, it does not allow for the essential distinction between different combinations of products containing similar groups. As an example, consider a matrix containing both SiH and SiCl, together with other products. The two infrared bands due to the species above could be assigned equally well to the two stretching vibrations of SiHCl. This molecule should also have a deformation vibration, but in the absence of any information from other sources about its frequency it would be possible to make a plausible assignment using a band anywhere in the range 1500–500 cm^{-1}. Such a mistake is unlikely in this instance, as all three molecules have been identified in gas-phase studies, but similar situations may arise in any matrix study where more than one species may be present.

Intensity variation method. An important safeguard against such errors is to ensure that all bands assigned to a single species vary in intensity in the same way as conditions are altered. The alterations possible will depend on the method used to form the species, but may include changes in duration and frequency of photolytic radiation, changes in the temperature of the matrix (which may allow formation or destruction of the species by diffusion-controlled reactions), alteration of reagent concentrations and of matrix ratios and materials. Several separate experiments will often be necessary to ensure an adequate range of variations in conditions.

The importance of this method has resulted in an insistence, unusual among non-analytical users of infrared spectrometers, on the measurement of the intensities (optical densities) of peaks in matrix studies. If this is done carefully it is possible to distinguish those that vary in the same way from all others. As matrix-isolated samples usually give sharp bands even closely adjacent peaks can be measured accurately. It is found in most cases that not only can bands due to product species (which increase as the amount of reaction increases) be distinguished from those due to precursors (which decrease as the reaction proceeds), but that intermediate products and by-products can also be distinguished from each other and from the final products. A useful distinction between products can also be made on the basis of the rate of destruction as the matrix warms up into the diffusion temperature range, large species being lost more slowly than small ones.

Isotopic substitution method. A most powerful method for the assignment of vibrational bands involves the use of isotopically substituted precursors. These will give rise to short-lived species that contain isotopes of different mass and hence have different vibration frequencies. Any bands that do not shift when a given atom is isotopically substituted must arise from species that do not contain that atom. The limiting factor on the usefulness of the method arises from the fact that in large molecules some vibration frequencies may be affected only very slightly by a change in mass of a particular atom, so the method is most useful for small species, where any isotopic change will alter all the vibration frequencies detectably. The sharpness of the bands produced by matrix-isolated species makes it feasible to measure shifts of the order of 1 cm^{-1}, or even less, so that in small molecules the effects of isotopic substitution may be detectable on all bands. The magnitudes of the changes in frequency of the various bands show which atoms are principally involved in each vibration.

To take a simple example, photolysis of methyl azide in an argon matrix at 4 K eventually produces a set of bands at 3620 cm^{-1}, 2029 cm^{-1} and 474 cm^{-1} whose intensities vary together with the time of photolysis. Using CD_3N_3 as precursor shifts the first and third bands most, showing that these bands correspond to vibrations in which a hydrogen or deuterium atom moves. ^{13}C substitution or ^{15}N substitution in the precursor shifts the second band significantly, confirming that this corresponds to a C≡N stretch, as might have been proposed on the grounds of group frequency correlations. The fact that ^{15}N substitution changes the first band position much more than ^{13}C substitution suggests that the corresponding vibration is an N–H stretch rather than a C–H stretch. These observations are consistent with the structure H—N≡C for the product of the photolysis. A further band, at 2046 cm^{-1}, which varies in intensity with the time of photolysis in a different fashion, does not shift on H/D substitution, and is assigned to CN, perhaps produced by photolysis of HNC.

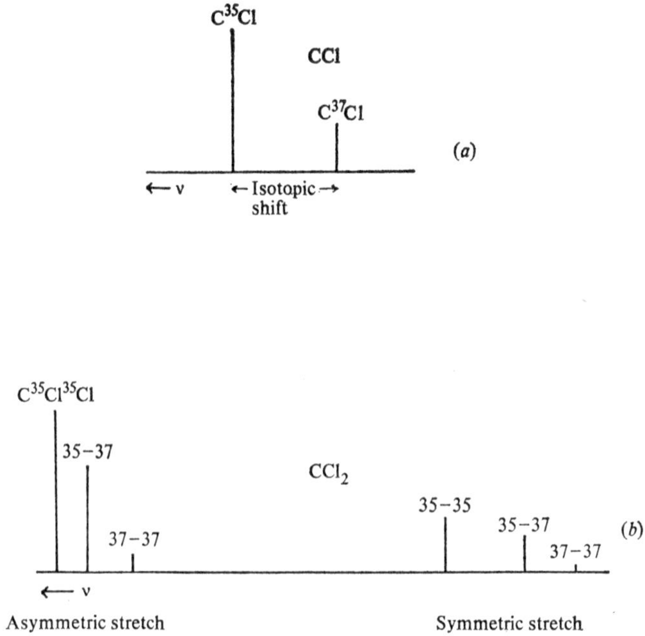

Fig. 5.3. Isotopic splitting patterns of (*a*) CCl and (*b*) CCl_2.

Isotopic splitting patterns. The use of isotope mixtures can lead to important extra information. To see why, we may consider the species CCl and CCl_2. The naturally occurring mixture of ^{35}Cl and ^{37}Cl will give rise to two distinct bands for CCl (see fig. 5.3(*a*)), corresponding to the stretching frequencies of the isotopic molecules $C^{35}Cl$ and $C^{37}Cl$. In the case of CCl_2 we may expect two CCl stretching modes from each species, but there will be three distinct species, $C^{35}Cl_2$, $C^{35}Cl^{37}Cl$ and $C^{37}Cl_2$. We then expect six bands in the spectrum (fig. 5.3(*b*)). If, as often happens, one of a pair of modes gives rise to a much more intense band than the other, the isotope pattern will be easily picked out, showing that the three components will be related to the ratio of isotopic abundances:

$$^{35}Cl:^{37}Cl = 3:1; \quad ^{35}Cl_2:^{35}Cl^{37}Cl:^{37}Cl_2 = 9:6:1.$$

This last set of ratios is far from ideal, as the weakest component may not be observable, especially if other bands are present in the same region of the spectrum. It is usually preferable to have equal proportions of the two isotopes, which can only be achieved artifici-

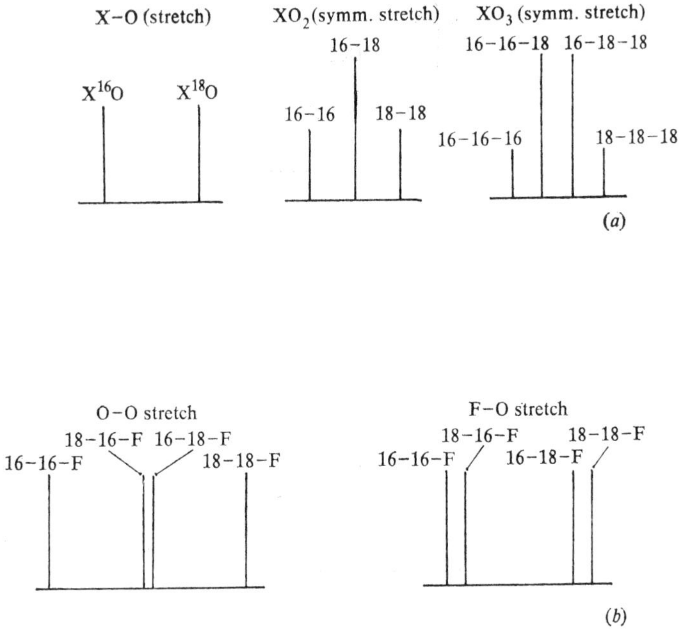

Fig. 5.4. Effects of 50 % ^{18}O enrichment in oxygen-containing molecules: (*a*) equivalent oxygen atoms; (*b*) non-equivalent oxygen atoms, e.g. O_2F.

ally by mixing in most cases. If this is done, species with one atom of the element concerned give 1:1 doublets, species with two atoms in equivalent positions 1:2:1 triplets, species with three equivalent atoms 1:3:3:1 quartets and so on, provided that the isotopes are randomly distributed (see fig. 5.4(*a*)).

If the positions are not equivalent, as in FO_2, for example, we find that in a randomly-mixed system containing ^{16}O and ^{18}O in equal proportions the four species $F^{16}O^{16}O$, $F^{16}O^{18}O$, $F^{18}O^{16}O$ and $F^{18}O^{18}O$ are present in equal amounts, so a 1:1:1:1 quartet results instead of a 1:2:1 triplet. The positions of the bands are not, of course, equally spaced (see fig. 5.4(*b*)). Thus if the F–O stretch of FO_2 is observed, we may expect to find two closely spaced bands due to the F–^{16}O stretches in $F^{16}O^{16}O$ and $F^{16}O^{18}O$, and two other closely spaced bands due to the F–^{18}O stretches in $F^{18}O^{16}O$ and $F^{18}O^{18}O$. By contrast, the bands of the 1:2:1 triplet for CCl_2 are expected to be roughly equally spaced; the effect of the added mass is cumulative here because the two chlorine atoms move equally in the vibrations.

The FO_2 system also emphasises the importance of random mixing of the isotopes. If F atoms are allowed to react with an equimolar mixture of $^{16}O_2$ and $^{18}O_2$ in a matrix, the product FO_2 contains only $F^{16}O^{16}O$ and $F^{18}O^{18}O$, as the O_2 molecule remains intact during the reaction. The expected 1:1:1:1 quartet is only obtained if the mixture of $^{16}O_2$ and $^{18}O_2$ is subjected to an electric discharge before the matrix is prepared. This treatment atomises the oxygen and recombination eventually leads to the equilibrium mixture of $^{16}O_2$, $^{16}O^{18}O$ and $^{18}O_2$ in the ratio 1:2:1.

Finally, it must be remembered that for some elements (notably fluorine and phosphorus) only one isotope is normally available or sufficiently stable for use in this way. In such cases information must be collected by isotopic substitution of the other atoms present to establish the probable structure of the species. The presence of fluorine or phosphorus must then be inferred from the spectrum as a whole, or from group frequency arguments. In some instances information may be obtained by replacing a fluorine atom by chlorine, but the resultant difference in bonding may make it very difficult to be sure of the correct correlation.

Normal coordinate analysis. This is the most sophisticated method used to establish the assignment of observed infrared bands. It involves computation, from assumed structures and force constants, of theoretical spectra which are then brought into coincidence with the observed bands by adjustment of the assumed structure or force constants. (Fig. 5.5 shows the major force constants for a molecule ABC.) Unfortunately, no unique set of force constants can be derived in this way for real systems without the use of other information to 'constrain' the solution so that it fits the observed spectrum *and* the additional information. One important source of additional data is provided by the spectra of isotopically substituted molecules. These provide more experimental data, so that the normal coordinate analysis can be repeated with only the atomic masses changed – the computation should then reproduce the isotopic spectrum.

Another source of extra information is the assumption of 'reasonable' values or limits for major force constants, or of 'reasonable' values for minor force constants, which indeed are often taken to be zero. There is, however, no general agreement on such reasonable values, and it seems philosophically unsound to operate on this basis. Indeed the assumption of 'reasonable' values for force constants has often lead to erroneous conclusions about structures. In any case,

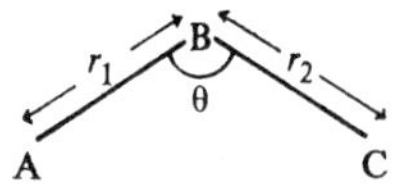

Fig. 5.5. Major force constants for the molecule ABC. If the force constant describing the A–B stretch is f_{r_1}, that for the B–C stretch is f_{r_2} and that for the ABC bend is f_θ, then the potential energy (V) of the vibrating molecule in terms of these force constants and bond/angle displacements from equilibrium is given by

$$2V = f_{r_1}\Delta r_1^2 + f_{r_2}\Delta r_2^2 + f_\theta \Delta\theta^2.$$

normal coordinate analysis is really only applicable after at least most of the spectrum of a particular species has been picked out, though it can be helpful in suggesting frequencies for bands that have not yet been observed or conclusively identified.

Perhaps the most satisfactory use of normal coordinate analysis is for the explanation of 'anomalous' isotope effects. These sometimes arise in a molecule with more than two atoms because the 'normal vibrations' are not necessarily the same as the idealised bond stretching vibrations and deformations, so that the isotopic substitution of one atom may affect more than one vibration frequency. If several vibrations of the same symmetry lie close together in energy, there may well be considerable mixing between them, which will be reflected in a 'sharing' of the expected isotope shifts among all the bands. This will be reproduced in the normal coordinate calculations if appropriate force constants are used; the extent of the mixing of the ideal modes can then be calculated.

The most serious defect of the method, apart from the necessity of making use of other information if a unique solution is to be found, lies in the assumption of purely harmonic motion. As anharmonicity effects are often of the order of 1 % of the observed frequencies it is pointless to try to reproduce the spectrum more precisely than this unless enough information on anharmonicity can be obtained to enable corrections to be made to the observed frequencies. This, of course, requires a far greater level of confidence in the assignment of the spectrum than is appropriate to our present purpose. Isotopic splittings, on the other hand, can be calculated with rather more confidence as the anharmonicity effects on the isotope bands tend to cancel each other.

Fig. 5.6 summarises schematically a typical characterisation route for a matrix-isolated system.

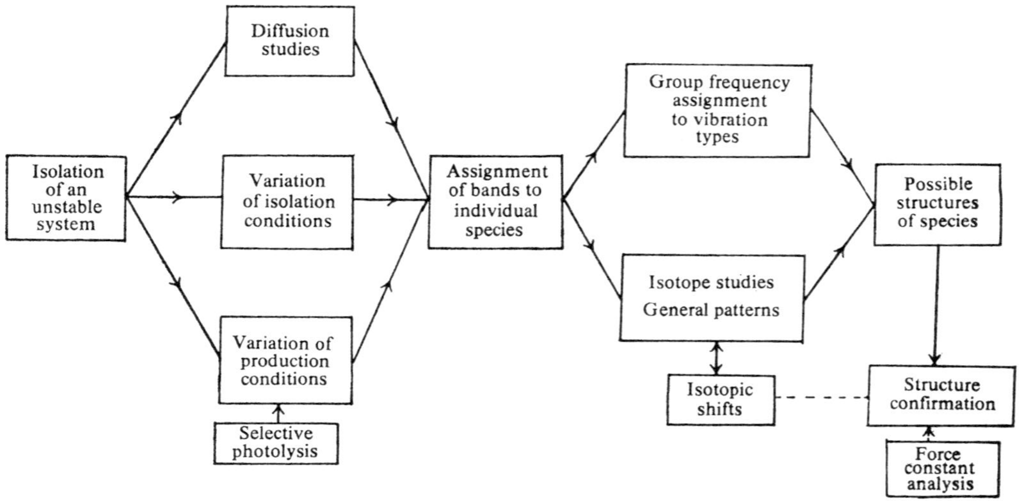

Fig. 5.6. Typical characterisation route for matrix-isolated species.

5.4 Electron spin resonance

This depends on the absorption of energy at radiofrequencies by samples that contain unpaired electrons in a magnetic field. It is inherently extremely sensitive, samples containing less than 10^8 unpaired electrons (less than 10^{-16} moles) giving detectable signals. Samples are usually much more concentrated than this, so the criterion for detection of an e.s.r. signal is simply that a species is present that contains one or more unpaired electrons.

Detection does not, of course, necessarily imply identification, and it is this step that makes e.s.r. relatively of minor importance in matrix studies. There is little 'shift' between the signals from different species, unless transition metal atoms with significant spin–orbit coupling effects are involved, so a single peak is of no use for identification. Essentially all the information that can be derived about the species responsible for a signal comes from the coupling of the electron spin with nuclear spins, which produces multiplets analogous to those arising from nuclear spin–nuclear spin coupling in n.m.r. spectra. Thus a pair of equivalent nuclei of spin $I = \frac{1}{2}$ give a 1:2:1 triplet in an e.s.r. spectrum, while two non-equivalent spin-$\frac{1}{2}$ nuclei give two doublets of equal intensity. As we are observing the electron spin resonance signal all couplings with nuclear spins are first order.

Couplings from different sets of nuclei superimpose to give complex multiplets; signals with hundreds of lines can be resolved in some cases. The effects of nuclear quadrupole moments, which can effectively decouple the nuclei concerned in n.m.r. spectra, are less drastic here, and even nuclei with large spins give rise to observable splittings. The coupling constants vary, for a given nuclear species, with the 'density' of the unpaired electron at the atom concerned; the relative sizes of the couplings due to different nuclei are determined by their relative gyromagnetic ratios.

The technique then can be useful only if nuclei with spin are present and coupled to the electron spin. Atoms without isotopes with nuclear spins can only be inferred from the spectra if they control the structure of the species, and hence the couplings from the 'active' nuclei. In some cases isotopic substitution is useful, as with oxygen, where the ^{17}O isotope has a nuclear spin; in carbon-containing species it is often possible to detect signals due to the 1 % of naturally-occurring ^{13}C.

A further important limitation is that, because of the lack of 'chemical shifts', weak signals due to minor components of a sample may be masked by signals from more abundant species. As it is rare for precursors to contain unpaired electrons there is not usually any signal from the precursor.

Practically, e.s.r. spectroscopy requires the sample to be accurately positioned, still of course at very low temperature and in a high vacuum, between the poles of an electromagnet in a resonant cavity tuned to a particular radiofrequency. The spectrum is scanned by changing the field strength of the electromagnet. One successful arrangement involves the use of a synthetic sapphire (Al_2O_3) rod, which has high strength and thermal conductivity at very low temperatures, and gives no e.s.r. signal if pure. This is cooled to the desired temperature and the matrix deposited on its surface, outside the resonant cavity, and the cooled rod, together with the refrigerant or refrigerator, lowered so that the rod and matrix enter the cavity. This allows trapping, co-condensation, irradiation and so on to be carried out away from the restrictions imposed by the magnet and the resonant cavity (see fig. 1.1(*c*)).

5.5 Raman spectroscopy

Like infrared spectroscopy, this involves vibrational excitation in most cases, but even homonuclear diatomic molecules give Raman spectra. As it is a non-resonant process the effect is weak and very intense light sources (lasers) and very efficient spectrometers are necessary for use with the very dilute samples needed for matrix isolation. Even so, most published matrix isolation Raman spectra show signals only from the major components of the sample, and the technique is most useful for the study of vibrations that give rise to no infrared bands (for symmetry reasons) in molecules that have been identified more or less conclusively by infrared methods. The frequencies of such vibrations are often important in normal coordinate analysis (see above). The experimental arrangement may be similar to that used for the study of electronic emission spectra (see fig. 1.1(*b*)).

6 Effects of the matrix on spectroscopic properties

Matrix studies rely on spectroscopic data to establish the nature of the species trapped and we have considered in chapter 5 how this may be done. It is necessary to remember, though, that the molecules we are studying are not in fact isolated as in the gas phase, and that the observed properties may be perturbed by the matrix environment. In this chapter we consider some of the effects of such perturbations. This will enable us to proceed to a consideration of the spectra observed in some experiments, and to conclusions about the nature of the species present.

From a very simple point of view, any spectrum is likely to consist of a number of peaks of particular position and shape. If we take the gas-phase (truly isolated) spectrum for a species as the standard, the spectrum of the matrix-isolated species may differ from this in respect of

(i) the number of peaks,
(ii) the positions of the peaks,
(iii) the shapes of the peaks.

The most commonly observed effect under heading (i) is the appearance of several closely spaced peaks where only one is expected, generally described as *matrix splitting*. There are several possible explanations for this, which we shall discuss shortly, but the most usual one is that several possible sites for the species exist in the matrix. The distinct peaks then arise from molecules trapped in the various distinct sites. This effect is clearly connected with (ii), the effect of the matrix on the peak position, generally known as the *matrix shift*.

Changes under heading (iii) vary considerably between the different spectroscopic techniques. Vibrational peaks in the infrared spectrum tend to be very much narrower in the matrix than in the gas phase because of the loss of rotational fine-structure, whereas electronic transitions in general give broader peaks in the matrix because of the influence of the matrix cage. Some e.s.r. signals seem to be rather little affected, but others are so much broadened as to be unobserved.

We must, then, consider in more detail the effects found in the various types of spectra in terms of the molecular processes associated with each, and the effect of the matrix upon such processes. The three most commonly used spectroscopic techniques are vibrational spectroscopy (infrared and Raman), electronic spectroscopy (visible and ultraviolet) and electron spin resonance spectroscopy. The last technique, which involves changes of electron spin in a magnetic field, must be treated separately, but the vibrational and electronic spectroscopic techniques, which involve changes in the vibrational and/or the electronic state of the molecule, may be treated together. The exceptional case of the electronic spectra of atoms, where no vibrational changes can occur and electron spin is important, must also be treated separately.

6.1 Effects of the matrix on electronic spectra of atoms

The main spectroscopic techniques available for the detection and study of atoms in matrices are electronic spectroscopy and e.s.r. spectroscopy. The electronic spectra of gas-phase atoms are very fully characterised, so electronic spectroscopy is in principle more useful than e.s.r., which is less easy to carry out for gaseous samples. Studies of the electronic spectra of matrix-isolated atoms in the visible and ultraviolet regions have shown that the sharp lines characteristic of the spectra of gas-phase atoms tend to be:

(i) broadened considerably (commonly by 100–500 cm^{-1}),

(ii) shifted significantly (by over 1000 cm^{-1} in some cases), usually to higher energies, but sometimes to lower, and

(iii) split into two or more components.

The breadth of the observed lines makes it difficult to measure their positions precisely, while the large shifts in position may make it difficult, especially in complex spectra, to assign correctly the bands to transitions observed in the gas phase.

The breadth of the electronic bands is at first sight rather surprising, as both infrared bands and e.s.r. bands are often narrow in matrix spectra. The effect presumably arises from perturbations due to overlap of the charge clouds of surrounding matrix atoms with the diffuse outer orbitals occupied in the excited state of the atom. These are not, of course, involved in vibrational or e.s.r. transitions.

The shift in position of electronic bands is likewise associated with overlap with the charge clouds of matrix atoms, and is ascribed to the effect of changes in effective size of the atom on excitation. Most metal atoms are larger than matrix atoms, but occupy single substitutional sites (see chapter 2) in annealed matrices, and are therefore 'squeezed' or subject to a repulsive potential. If the excited atom is effectively larger than the ground-state atom the increased potential energy due to the repulsive force acting on the larger excited atom will increase the transition energy.

A decrease in the transition energy from the gas-phase value can be expected if the atom is in an attractive potential, such as might arise if the effective size of the atom were less than that of the site occupied. This is more likely to occur for large matrix atoms, such as xenon, and small trapped atoms.

The spectra of the alkali metal and coinage metal atoms (lithium, sodium, potassium, rubidium, caesium, copper, silver and gold) are in principle simple and illustrate well the effects we are discussing here. In particular, they show the splittings referred to above in a dramatic fashion. The transition most thoroughly studied is the ns^1–np^1 excitation in each case.

In the gas phase, this excitation gives rise to two lines, which correspond to the two possible electronic states of the atom with a single valence-shell p-electron, $^2P_{\frac{3}{2}}$ and $^2P_{\frac{1}{2}}$. The splitting is associated with spin–orbit coupling in the excited state of the atom. This coupling increases rapidly as we go from lithium to caesium or from copper to gold, and gives splittings ranging from 0.3 cm^{-1} for lithium to 550 cm^{-1} for caesium and from 250 cm^{-1} for copper to nearly 4000 cm^{-1} for gold.

In the matrix, on the other hand (see fig. 6.1) all the atoms mentioned give *three* broad bands in the region where these transitions are expected. The breadth of the bands makes it very difficult to specify their positions, and there is little agreement between the values given in the many different reports. It seems clear, though, that the three bands are all genuine, all associated with isolated metal atoms in a single type of site and all derived from the ns^1–np^1 transition.

For silver and gold the three bands are unequally spaced, one being separated from the other two by about 1000 and 4000 cm^{-1} respectively. This suggests (see below) that the isolated band in these cases corresponds to the $^2P_{\frac{1}{2}}$ state of the excited atom, and the other two bands in each case to the $^2P_{\frac{3}{2}}$ state. This comparatively simple assignment is not possible in the spectra of the other metals, where the three bands are separated in all cases by some 300–600 cm^{-1}, and no

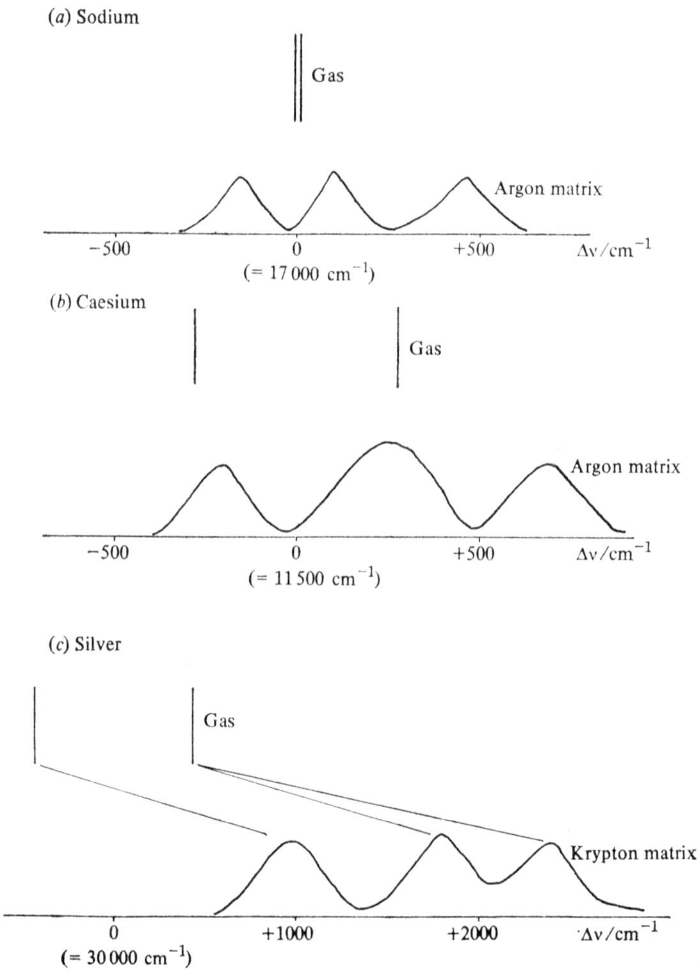

Fig. 6.1. $ns^1 \rightarrow np^1$ bands for sodium, caesium and silver in the gas phase and in matrices.

splittings corresponding to the spin–orbit splittings observed in the gas phase can be detected. For the lighter metals some other splitting effect is clearly predominating over the spin–orbit coupling.

This must be due to the effect of the matrix on the ^2P upper state of the atom. The usual single substitutional site in a close-packed matrix has octahedral symmetry, which can only give rise to two components of a ^2P state (analogous to the two components that arise from the spin–orbit interaction in a gas-phase atom). One of these components, however, is still *orbitally degenerate*, and this orbital degeneracy will be removed only by a distortion of the cage that lowers the symmetry about the atom. The distortion may be either tetragonal or trigonal, depending on whether it occurs by an elongation along one of the four-fold axes or one of the three-fold axes of the 12-atom cage. (The effect is entirely analogous to the Jahn–Teller distortion of an octahedral transition-metal complex with an orbitally degenerate electronic state.) The overall effect of the matrix cage is then to remove completely the original three-fold degeneracy of the p-orbitals, giving the three components of the transition as observed.

This sort of effect must be expected to occur whenever an atom is present in a matrix in a state with orbital degeneracy (orbital quantum number $l \neq 0$). It should therefore affect *all* electronic transitions, for which a change in l is involved. Where the ground state of the atom has $l \neq 0$ the splitting will affect the e.s.r. spectrum as well, and this probably accounts for the fact that no e.s.r. spectra of atoms such as the halogens, with ^2P ground states, have been reported.

6.2 Electron spin resonance (e.s.r.) spectra

The two main parameters associated with an e.s.r. spectrum are the g-value and the coupling constants with any magnetic nuclei. A matrix may in principle induce changes in either or both of these, and may also introduce additional couplings if the matrix atoms themselves have magnetic nuclei.

Studies of atoms and molecules show that matrix shifts in the g-value are small where g is near the 'spin-only' value of 2.0. If this is not so, in cases where spin–orbit coupling effects are important, considerable shifts may occur as the matrix changes the effective spin–orbit coupling.

More generally important are changes in nuclear–electron coupling constants, which are relatively small for hydrogen, one of the best-studied atoms, but considerably larger for atoms such as nitrogen with occupied p-orbitals. Similar changes are reported for molecules; they are ascribed in this case to changes in hybridisation associated with the influence of the matrix upon bond angles.

Coupling to matrix atom nuclei has been observed for metal atoms in matrices of krypton and xenon, both of which have some magnetic nuclei (11.5 % ^{83}Kr, $I = \frac{9}{2}$; 26.4 % ^{129}Xe, $I = \frac{1}{2}$; 21.2 % ^{131}Xe, $I = \frac{3}{2}$). The observations are consistent with, but insufficient to prove, the presence of 12 equivalent matrix atoms around each metal atom. Such coupling is small because of the relatively large non-bonded distance between the atoms and the lack of significant involvement of the closed electron shells of the matrix atoms with the unpaired electrons of the metal atoms.

It seems unlikely that electron–electron coupling between non-nearest-neighbour trapped atoms will give rise to complex e.s.r. spectra; instead, spin-pairing will probably occur even over rather large distances and the e.s.r. spectrum will be lost. It is possible that some of the splittings observed for hydrogen and alkali metal atoms arise from very long-range couplings of this type, but no proof of this has been found. Such splittings are more likely to arise from the presence of atoms trapped in a variety of different sites, some substitutional, some interstitial, and some perturbed by the presence of impurities (such as precursor molecules) in or near the immediate cage.

6.3 The effects of the matrix on the vibrational and electronic spectra of molecules

Early studies showed that changes in (*a*) the number, (*b*) the position, and (*c*) the shapes of peaks could be observed in the spectra of molecules isolated in matrices, as compared with their spectra in the gas phase. While some of these effects, such as the elimination from absorption spectra of bands due to absorption by molecules in excited states, are easy to understand, other effects are more complex. It becomes necessary to consider in some detail how the interaction of the matrix with a trapped species can affect the observed spectroscopic properties. We begin by treating the case of a diatomic molecule.

For an idealised harmonic oscillator the appropriate potential energy curve is a parabola (fig. 6.2), the potential energy V being given by

$$V = \tfrac{1}{2}k(r-r_e)^2.$$

In this idealised case the quantised vibration energies permitted are given by

$$G_v = \omega(v+\tfrac{1}{2}),$$

where v is an integer, the vibrational quantum number, and ω (omega) is the classical vibration frequency, related to the force constant, k, and the reduced mass of the system μ by

$$\omega^2 = \frac{k}{4\pi^2\mu}.$$

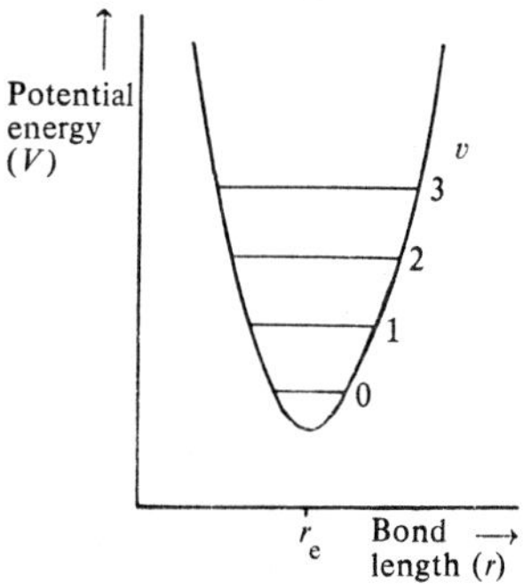

Fig. 6.2. Potential energy curve of harmonic oscillator.

The permitted energy levels are equally spaced, and are separated by ω. In the lowest possible state, the vibrational ground state, the molecule retains a *zero-point energy* of $\frac{1}{2}\omega$.

In real systems we must take account of the anharmonic nature of the oscillations. This arises because the potential energy curve for a real diatomic molecule is not a parabola; it rises more steeply than the parabola for $r < r_e$, and levels off at the dissociation energy as $r \to \infty$. Such a curve (see fig. 6.3) is not readily expressed by a formula, and the calculation of exact expressions for the allowed energy levels from such a formula is impossible. It is therefore customary to treat the anharmonic oscillator as a perturbed harmonic oscillator, and to express the perturbed set of permitted energy levels by

$$G_v = \omega_e(v+\tfrac{1}{2}) - \omega_e x_e(v+\tfrac{1}{2})^2 + \omega_e y_e(v+\tfrac{1}{2})^3 - \dots$$

Only the first two terms are usually retained; ω_e is then the so-called harmonic vibration frequency of the system, and $\omega_e x_e$ the anharmonicity. It should, however, be apparent that neither has any physical significance on its own; together they may be used to generate the set of vibrational energy levels, or the set of observed vibrational transitions between these levels.

Electronic transitions occur between two states of a molecule, each of which has its own potential curve. If the curve has a minimum there will be quantised vibrational levels, and transitions to more than one of these may be observed. In such a case information about the vibrational energy levels may be derived from the electronic spectrum. An absorption spectrum will, for a matrix-isolated sample,

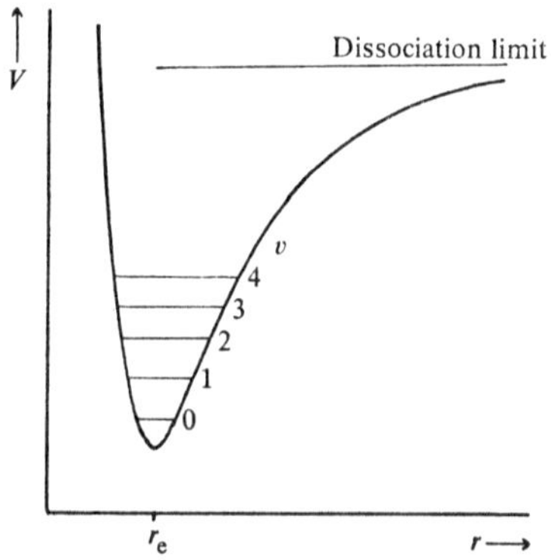

Fig. 6.3. Potential energy curve of anharmonic oscillator.

contain transitions from the vibrational ground state of the lower state to various vibrational levels of the electronically excited state, and thus give information about the upper potential curve, while an emission spectrum will usually occur from the lowest vibrational level of the upper state to various vibrational levels of the lower electronic state and give information about the lower potential curve. In each case the observed band will give information about the electronic transition energy, the difference between the two $v = 0$ levels, and about the vibrational *term values* for one of the states, given by

$$G_v - G_0 = v\omega - v(v+1)\,\omega x.$$

Potential curves of a matrix-isolated diatomic molecule. The effect of the matrix on the energy levels, and hence on the observed vibrational and electronic transition energies, of a diatomic molecule may similarly be treated as a perturbation of the harmonic potential by the interaction of the vibrating molecule with the cage that contains it. If we treat the cage as effectively a sphere of diameter D (fig. 6.4) it seems reasonable to add to the potential energy of the molecule a term $b(D-r)^{-n}$ to express the repulsive force between the cage and the molecule, as a function of the internuclear distance r of the molecule. The exponent, n, will usually be in the range 6–12, as for other cases involving interatomic repulsion.

The modified potential energy is now given by

$$V' = \tfrac{1}{2}k(r-r_e)^2 + b(D-r)^{-n}.$$

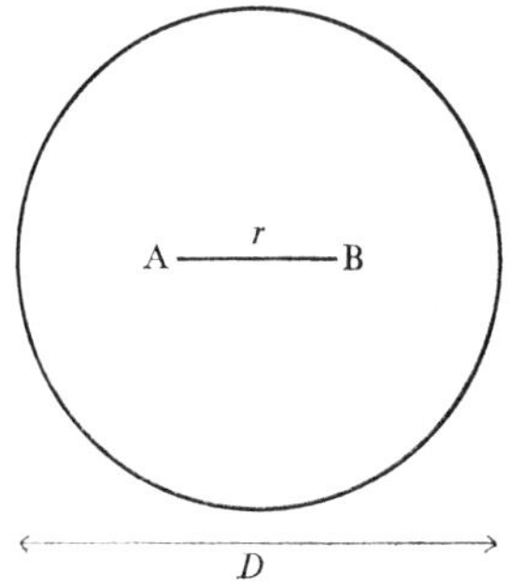

Fig. 6.4. Diatomic molecule in a spherical matrix cage.

The potential minimum, or equilibrium position, occurs when $\mathrm{d}V'/\mathrm{d}r = 0$, that is when

$$k(r-r_e)+nb(D-r)^{-(n+1)} = 0;$$

or
$$r'_e = r_e - \frac{nb}{k}(D - r'_e)^{-(n+1)}.$$

The equilibrium bond distance (now r'_e) has thus *decreased.*

The potential energy of the minimum has *risen* to

$$V'_e = \frac{n^2b^2}{2k}(D-r'_e)^{-2(n+1)}+b(D-r'_e)^{-n}.$$

The third effect of added repulsive potential upon the vibration frequency is best seen by calculating the new effective force constant at the new potential minimum. Just as

$$\frac{\mathrm{d}^2V}{\mathrm{d}r^2} = k$$

so
$$\frac{\mathrm{d}^2V'}{\mathrm{d}r^2} = k+n(n+1)b(D-r'_e)^{-(n+2)}.$$

The effective force constant is thus increased, and the vibration frequency will *increase* because it is directly proportional to the square root of the force constant.

The effect of a repulsive potential on an anharmonic oscillator will, of course, be similar to the effect on a harmonic oscillator. In addition, since the additional repulsive potential makes the total potential energy curve more nearly parabolic (by, particularly, counteracting the tendency to dissociation), we may expect the matrix to *reduce* the anharmonicity (see fig. 6.5).

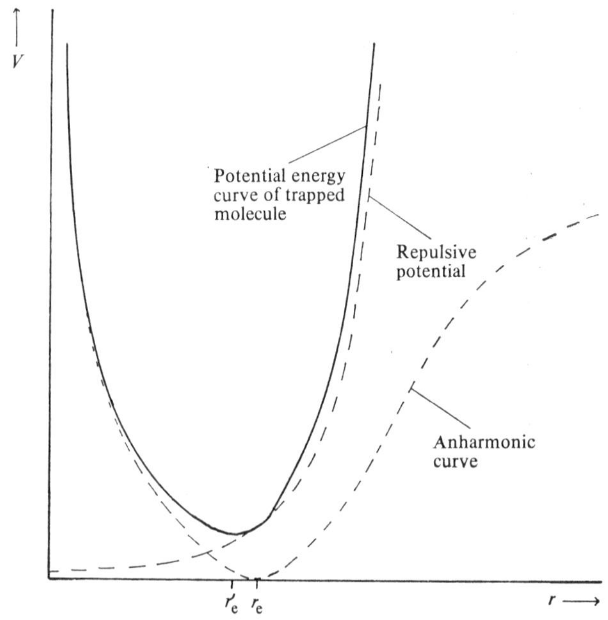

Fig. 6.5. Effect of repulsive potential on anharmonic potential energy of a trapped diatomic.

Conversely, an attractive potential between the trapped molecule and the cage will have the effects of

(i) increasing the equilibrium bond distance,
(ii) lowering the potential energy of the minimum,
(iii) decreasing the vibration frequency, and
(iv) increasing the anharmonicity.

Examples of both types of potential have been found, and in some cases the effects studied for different matrices. Where electronic transitions are involved it is often found that the effect of the matrix changes, being attractive for one state and repulsive for the other. Such changes can be correlated with the expected changes in equilibrium bond length between the two states.

The effect of the matrix on an electronic transition energy, or band position, is of course particularly complex. Not only does the shift from the gas-phase value involve the *difference* in the changes in potential energy of the minima of the potential curves for the two states, but also the difference in the changes in the vibrational zero-point energy for the two states. The latter factor is usually comparatively small, but the relative shift in position of the two potential curves can be quite large. Elementary analysis suggests that when a bonding electron is excited the bond length will tend to increase during the transition; the matrix potential is therefore likely to be more repulsive in the upper state, and the upper state will be raised in energy more than the lower. The overall transition energy should thus increase in the matrix, giving a shift to shorter wavelength (a 'blue-shift'). Conversely, excitation of an antibonding electron should tend to reduce the bond length, and the effect of the matrix is likely to reduce the transition energy, giving a red-shift.

Polyatomic molecules. The situation for a *polyatomic molecule* is, of course, more complex than for a diatomic, but it seems reasonable to retain the concepts of repulsive and attractive matrix potentials superimposed on the molecular potential. An increased vibration frequency is then said to be due to a repulsive matrix cage, a decreased frequency to an attractive matrix cage.

Interestingly, it is found that even for the same molecule in the same matrix some bands in the vibrational spectrum may increase in energy while others decrease in energy. A diagram (fig. 6.6) showing the shifts found for bands of a number of compounds as a function of frequency shows that for bands below 1000 cm^{-1} shifts are most often positive (increased frequency in the matrix), whereas for bands above 1000 cm^{-1} shifts are usually negative. Where opposite shifts are found for different bands of the same molecule it is necessary to suppose that one vibration takes place in the presence of an effective attractive potential due to the matrix, while the other is subject to a repulsive potential. When the two vibrations involve motion in different spatial directions it is easy to visualise how this may arise, but it seems that even two vibrations involving motions in the same direction may show opposite shifts. In such cases it may be supposed that the high energy vibration is somehow able to influence the shape and size of the cage so that the effective potential is attractive, whereas the low frequency vibration is subject to the normal repulsive effect of neighbouring matrix atoms. It may be noted that high energy vibrations contribute most to the zero-point energy of the system, which is a property of the trapped molecule as such. Any reduction in the high energy vibration frequency due to an attractive potential contributes directly to a lowering of the zero-point energy and therefore a reduction of the total energy of the system even in its ground state.

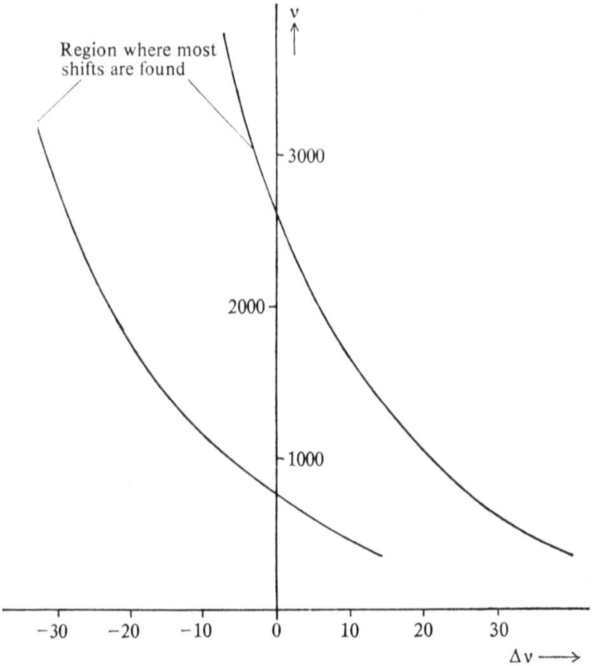

Fig. 6.6. Variation of matrix shift with frequency.

The electronic spectra of polyatomic molecules are subject to shifts in a matrix, and are usually found to give somewhat broadened bands, as for diatomics. The shifts can often be rationalised in terms similar to those used for diatomic molecules. As there are several fundamental vibration frequencies that may be excited in most cases, it is often difficult to analyse the vibration structure of electronic bands of polyatomic molecules because of overlapping of the broad vibrational peaks.

6.4 Matrix splittings in vibrational and electronic spectra

We must now return to a consideration of the possible mechanisms that may be involved in the appearance of close groups of peaks in the vibrational and electronic spectra of matrix-isolated molecules, often called spectral multiplets. This phenomenon is particularly common in vibrational spectra, where each band is usually very sharp, so that peaks within 1 cm^{-1} of each other can readily be resolved given adequate instrumental resolution. A number of explanations have been suggested for this effect, and it seems likely that several of them may be important.

Relaxation of selection rules and lifting of degeneracy. These effects result from the lowering of symmetry from the isolated gas-phase molecule to the molecule trapped in a disordered matrix. It is possible in principle for all vibrations to be infrared active in the latter situation, so additional peaks may arise due to fundamental vibrations forbidden in the gas phase. There is no reason why such peaks should occur close to others, so the appearance of spectral multiplets for this reason is somewhat unlikely. More plausible is the suggestion that in the disordered matrix the degeneracy of vibrations may be lifted, giving rise to two components from a doubly degenerate vibration or three from a triply degenerate vibration. This definitely seems to occur in many cases, but is only possible for molecules with symmetry axes of order greater than two, as only such molecules can have strictly degenerate vibrations.

Coupling to lattice modes of the matrix. In some cases close multiplets have been assigned as due to coupling of the internal vibration of a molecule with low energy lattice vibrations of the matrix. Very little is known about the vibrations of disordered solids, but it seems unlikely that such coupling could give rise to sharp peaks, as the disorder of the matrix will give rise to rather broad bands of lattice modes.

Rotation and translation within the cage. It is now well established that for very small molecules, such as OH, NH_2 and NH_3, quantised rotation may be possible in some matrices. Thus NH_3 appears to rotate in an argon matrix but not in nitrogen. This manifests itself in the appearance of separate R(0), Q(1) and P(1) lines in each vibrational and electronic transition, corresponding to changes in the rotation quantum number J of $0 \rightarrow 1$, $1 \rightarrow 1$ and $1 \rightarrow 0$ respectively. In some instances transitions starting from higher rotational levels than $J = 1$ are observed if the circumstances of the experiment allow such levels to be continuously populated by absorption of energy followed by re-emission. A characteristic of multiplets due to rotation is that marked changes occur reversibly in their relative intensities as the temperature of the matrix changes because the thermal energy appropriate to the temperature (4–10 K) is similar to that needed to populate the $J = 1$ level from the $J = 0$ level.

The splittings observed for OH in rare gas matrices cannot be explained solely as due to rotation. It has been suggested that both rotation and quantised translation of the OH within its cage must be involved.

Multiple site effects. As we have seen, the matrix cage may shift a vibrational or electronic band from its gas-phase position; it is therefore possible that different types of matrix cage may induce different shifts, so that close multiplets, or wider splittings, may be due to the presence of molecules trapped in a variety of sites. The existence of such a variety of sites arises from the essentially random but non-equilibrium nature of the trapping process. If the matrix is kept below the annealing temperature during and after deposition the elimination of less favourable sites in favour of more stable ones is prevented.

The concept of multiple sites was first put forward to explain the observation of several types of hydrogen atoms trapped in matrices. This system probably involves both interstitial and substitutional sites. For larger molecules the various sites adopted will involve more or less perfect packing in the immediate cage and the surrounding layers, and the presence of impurities in or near the cage. All these may affect the effective matrix potential and hence the vibration frequency or the electronic transition energy of the molecule, giving rise to a variety of bands in the spectra.

An objection to this explanation is that the essentially random structure of the matrix would surely lead to a continuous range of possible sites, giving broad bands rather than sharp multiplets as observed. Broadish bands are in fact often obtained on initial deposition, but annealing of the matrix after deposition leads to the formation of sharp multiplets. This suggests that the various sites that give rise to the sharp peaks are either very similar in energy, so that no tendency exists for interconversion, or are very distinct from each other, so that interconversion is physically impossible. The most likely difference between sites that would be impossible to eradicate without complete mobilisation of the matrix is that they have cages containing different numbers or arrangements of impurities (e.g. nitrogen from air leaks). As we have suggested in chapter 3, it is likely that much of the published work with rare gas matrices in fact involved considerable impurity levels.

Aggregation and the effects of non-nearest neighbours. The existence of multiple sites, then, is quite likely closely related to the phenomena associated with interactions between close and distant neighbour molecules. As we showed in chapter 2, few matrix studies are carried out under conditions where the interactions of neighbouring molecules can be ignored, and it seems more than likely that most spectral multiplets can be accounted for by such effects.

The effect of aggregation upon a vibration frequency will, of course, depend on the strength of the interaction between the species concerned. In the limit of a weak interaction with a neighbour outside the immediate matrix cage only slight shifts in the vibration frequencies may be expected; a variety of such interactions could readily give rise to a multiplet. The effect of allowing diffusion to occur would be to diminish the intensity of the peaks due to such weakly interacting pairs, which would move closer together to form definite dimers.

Dimers may behave as two loosely linked monomer molecules, if only weak forces exist between the portions; in this case only small changes in frequency are expected. Stronger interactions, such as chemical bond formation, will result in the spectrum of the monomer being replaced by that of the dimer. In most cases dimer spectra may be distinguished from spectra due to monomers and loose aggregates because the proportion of dimers is likely to increase as diffusion proceeds, at the expense of monomers and loose aggregates.

Distinguishing multiplets due to different causes. This is best done by a study of the effects of varying the deposition conditions and the matrix temperature on the multiplet structure. A multiplet that is essentially unaltered so long as the matrix remains intact is likely to arise from some internal property of the molecule associated with the loss of symmetry in the matrix. Multiplets showing marked reversible changes as the temperature of the matrix alters are most likely due to rotation of a small molecule. In cases where the multiplet structure changes slowly and irreversibly as the matrix warms up, it is likely that an interaction with neighbouring impurities or other molecules is involved. Multiplets that are replaced by dimer absorptions following diffusion must arise from loose interaction between similar molecules. The most satisfactory way to establish that a multiplet arose from the presence of impurity species from air leaks would be to alter the deposition conditions so as to alter the proportions of impurities present – faster deposition is likely to lead to lower concentrations of impurities if the impurities collect during deposition. Splittings due to impurities in the sample itself should vary with the matrix ratio, while impurities in the matrix gas should be eliminated by using very high purity gases for the matrix studies.

7 Fragment molecules studied in matrices

In this chapter we shall examine the fragment molecules prepared by photolysis, atom transfer reactions in matrices or by discharge methods. It is convenient to divide the many small reactive species which fall into these categories into those containing no hydrogen and those containing one or more atoms of hydrogen. Within each section we shall consider species in order of increasing mass and complexity.

I. NON-HYDRIDE SPECIES

7.1 Atoms

A number of atoms have been studied in matrices and others are known to be formed by photolysis but have not yet been detected spectroscopically. No vibrational transitions are expected for atoms, though the presence of the atoms in the matrix may influence the frequency and activity of lattice modes of the matrix. Atoms are thus mainly detected and studied by e.s.r. or electronic spectroscopy.

The e.s.r. parameters of many atoms are known from gas-phase studies, usually involving atomic beam techniques. It is found in general that matrix effects do not alter the *g*-values much, but may have a substantial effect on the nuclear spin–electron spin coupling constants *A*.

Atoms such as the alkali metals, with ^2S ground states, give spectra showing the effects of:

(i) coupling of the electron spin to the nuclear spin *I* of the alkali metal atom;

(ii) coupling of the electron spin to the nuclear spins of the 12 surrounding matrix atoms (only in krypton and xenon matrices is this effect large enough to be detected);

(iii) in some cases the existence of more than one possible type of site in the matrix for the atom.

The process of annealing is usually found to remove the effects associated with multiple sites, indicating that one type of site is energetically favourable in the close-packed cubic lattice. This is probably a substitutional site with 12 equivalent nearest neighbours.

Interestingly, boron and halogen atoms (with ^2P ground states) do not seem to have been detected by e.s.r. As these atoms are *smaller* than the alkali metal atom it is unlikely that they would be unable to take up stable positions in the matrix. It seems more probable that the removal of the orbital degeneracy by distortion of the matrix site, which leads to separation and broadening of the electron energy levels (see chapter 6), is involved.

The group V atoms nitrogen, phosphorus and arsenic have been observed; the spectra show the expected couplings to nuclear spin. The coupling constant A for nitrogen atoms varies very substantially from one matrix to another. No complications due to the presence of three unpaired p-electrons (^4S states) are apparent.

The *electronic* (UV/visible) spectra of some matrix-isolated atoms have been studied in detail. As we discussed in chapter 6, splittings and shifts occur, and these may be analysed in terms of the potential exerted on the atom by the matrix and the nature of the transition involved. Metal atoms may be evaporated from ovens, or 'sputtered' by allowing a stream of matrix gas that has passed through a discharge to pass over a metal surface just before the gas is condensed. In general, only transitions (in absorption) arising from the atomic ground state are observed.

It is interesting, in passing, to note that no electronic spectra of non-metal atoms in matrices appear to have been reported. While this is doubtless partly explained by the higher ionisation potentials of non-metal atoms it may well be that the broadening and shifting of bands observed for metal atoms are greater for atoms with occupied p-levels due to the more diffuse and less symmetric nature of the p-orbitals.

7.2 Homonuclear diatomic molecules

Several homonuclear diatomic molecules have been trapped in matrices; we shall consider a metal Li_2; three members of a single group C_2, Si_2 and Pb_2; and a non-metal S_2. Again, no infrared spectrum is expected (the IR spectrum earlier reported as possibly due to matrix perturbed S_2 has now been shown to be due to the dimer S_4), and no e.s.r. spectra have been reported, though Si_2, Pb_2 and S_2 have triplet ground states.

Li_2, although present as < 1 % of the vapour from the evaporating oven, is the main species trapped by co-condensation of lithium with rare gases unless the matrix ratio is higher than 10000:1. It is suggested that dimerisation occurs during deposition in a fluid surface phase before freezing takes place. Similar dimers are formed by the other alkali metals at low matrix ratios, and even by mercury. In each case the species is identified only by the energy of the first electronic transition and the associated vibration frequency.

C_2 is produced in matrices by UV or X-ray photolysis of acetylene, or of methane; it is not certain that it can be detected when the vapours from evaporation of graphite are trapped. While the electronic ground state of C_2 is ${}^1\Sigma_g^+$, a low lying triplet state (${}^3\Pi_g$) exists that gives rise to the Swan bands in comets and in hydrocarbon flames. Early work on trapped C_2 suggested that *both* these states could be trapped in a matrix, and absorption spectra recorded. Later, however, it was shown that the bands assigned to the 'Swan' system of ${}^3\Pi_g$ C_2 were in fact due to C_2^-, produced in the matrix by photoelectron transfer. Thus only the ground state of C_2 is in fact observed in absorption, though the Swan bands can be detected in emission during the X-ray bombardment of acetylene in matrices. C_2^- was one of the first ions to be identified in a matrix; the number of ions produced during the photolysis of trapped acetylene increases if photoelectron sources, such as caesium or trimethylamine (with low ionisation potentials) are present. Irradiation of matrices containing C_2^- with light in the 200–280 nm region (about 5 eV) removes the absorptions attributed to C_2^-, which is consistent with the estimated ionisation potential (~ 4 eV). The processes

$$C_2 + e^- \rightarrow C_2^-$$

and

$$C_2^- \xrightarrow{h\nu} C_2 + e^- \quad h\nu > 4.5\ \text{eV}$$

can thus be studied and are effectively reversible.

Si_2 has been reported both from the photolysis of SiH_4 in a matrix and from trapping of the vapour from evaporation of solid silicon or silicon carbide. The ground state of Si_2 is ${}^3\Sigma_g^-$, unlike that of C_2, and the transition $H({}^3\Sigma_u^-) \leftarrow X({}^3\Sigma_g^-)$ occurs strongly in matrices at 390–350 nm (25000–29000 cm^{-1}); a vibrational progression ($\omega' \sim$ 260 cm^{-1}) correlates well with that found in the gas phase.

Experiments involving the evaporation of lead gave bands at 500 nm and 255 nm attributed to Pb_2 under conditions where surface diffusion during deposition was probable. The stretching frequencies in the upper state of the first band varied from 140 cm^{-1} to 180 cm^{-1} in different matrices, compared with the value of 159 cm^{-1} in the gas phase. The second band showed no vibrational structure; it is close to the atomic line corresponding to the $^3P(6s^26p^17s^1) \leftarrow {}^3P(6s^26p^2)$ transition of lead and presumably involves a similar excitation in the molecule.

S_2 has been trapped in matrices; the vapour from molten sulphur was heated to 1000 °C and then passed through a discharge to dissociate the larger species. Even so, IR absorptions due to S_4 were observed. The electronic transition $B(^3\Sigma_u) \longleftrightarrow X(^3\Sigma_g)$ was studied in detail both in absorption and in emission. The shifts in the 0–0 transition, the upper and lower vibration frequencies and the anharmonicities were measured for each of the rare gas matrices and accounted for in terms of the effects of the matrix on the potential curves of the ground and excited states.

7.3 Heteronuclear diatomics

Most of the heteronuclear diatomic molecules detected in matrices are metal monoxide or monohalides, which are discussed in chapter 8. Of those not containing metals, the majority contain only first row atoms. The best attested are BN (8 electrons), CN and BO (9e), CF (11e), NF (12e) and OF (13e). In addition BS, NCl, NBr and OCl have been reported.

BN, which is isoelectronic with C_2, has by contrast a $^3\Pi$ ground state, similar to the $X(^3\Pi_u)$ state of C_2 that is the lower state of the Swan system. It is produced by photolysis of $H_3B \leftarrow NH_3$ during matrix deposition; only the electronic spectrum has been reported.

CN, with one more electron, has a $^2\Sigma$ ground state, and the e.s.r., infrared and electronic spectra have all been reported. It is produced by photolysis of HCN in a matrix; this photolysis also gives the isomer HNC (see below). Photolysis of the cyanogen halides FCN, ClCN and BrCN gives the isomeric XNC but not CN itself, suggesting that the isomers may be formed by recombination after dissociation. The different behaviour of HCN is probably due to the mobility of hydrogen atoms in the matrix.

BO and BS have also been observed by e.s.r. They were trapped from vapours effusing from ovens containing solid boron with barium oxide or zinc sulphide.

7.4 Triatomic molecules

It is a well-known result of bonding theory that triatomic molecules with 16 valence electrons or less are expected to be linear, while those with 17 or more are expected to be bent. Several such reactive molecules have been trapped in matrices and the structures shown to be in accordance with the '16e rule'. Some restrictions to its validity are particularly well shown by matrix results.

Triatomic molecules (see table 7.1) containing only the first row elements boron, carbon, nitrogen, oxygen and fluorine obey the 16e rule. Thus BC_2 (11e), C_3 (12e), CCO, CNN and NCN (14e), NCO (15e) and CO_2, FCN and FNC (16e) are all linear. Their bending vibrations increase steadily in frequency, showing the influence of the increasing numbers of π-bonding electrons.

On the other hand, FCO and FOO and the difluorides BF_2, CF_2, NF_2 and OF_2 are all bent, having more than 16 valence electrons. This of course is also true of NO_2 and O_3. Other similar species containing one or more second row atoms (e.g. SiF_2, PF_2, ClF_2; ClCF; CCl_2, etc.) also have bent structures. It is of interest to note that unlike CH_2 the carbenes CF_2, ClCF and CCl_2 all have *singlet* ground states.

One interesting report suggests that Cl_3 has a *symmetrical* linear structure, which is perturbed by the effects of nearby Cl_2 molecules. The result is based on the observation of 8 components of a single stretching mode in the IR. Raman spectra of Br_3 suggests a symmetrical linear structure. Both these molecules were produced in a discharge before deposition of a mixture of the diatomic halogen and krypton.

Other exceptions to the 16e rule occur when ionic structures become plausible. Thus the lithium compounds LiON, LiO_2 and LiOF, with 12, 13 and 14 valence electrons, are all bent, LiO_2 (and the other alkali metal super-oxides) having a symmetric triangular structure 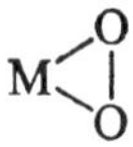 (see chapter 8). Again, the alkaline earth metal difluorides are bent in matrices, and it has been suggested that this is due to the dipole-induced dipole interaction of the bent species with neighbouring matrix atoms (the linear molecule of course has no dipole).

The species BF_2, NF_2 and PF_2 have each a single unpaired electron and have been detected by e.s.r. It is found that BF_2 and NF_2 are

TABLE 7.1. *Some triatomic molecules, involving first row elements, observed in matrix studies*

Molecule	No. of electrons	Linear or bent	Bending frequency ν_2/cm^{-1}
BC_2	11	L	
C_3	12	L	70
CCO	14	L	381
CNN	14	L	393
NCN	14	L	423
NCO	15	L	487
FCN	16	L	455
FNC	16	L	
CO_2	16	L	662
NO_2	17	B	750
FCO	17	B	626
BF_2	17	B	
CF_2	18	B	668
FOO	19	B	585
NF_2	19	B	
OF_2	20	B	461

effectively freely rotating in xenon or argon respectively at 4 K, though the rotation of NF_2 in neon can be 'frozen' at 4 K. PF_2, on the other hand, is effectively non-rotating in argon even at 20 K, though it rotates in xenon at higher temperatures. BF_2 was produced by γ-radiolysis of BF_3 in the matrix, NF_2 by trapping the partially dissociated vapour of N_2F_4 and PF_2 by γ-radiolysis of PF_3, by photolysis of PF_2H or P_2F_4 or by trapping from the vapour after pyrolysis of P_2F_4.

Triatomic ions. A few triatomic ions have been reported in rare gas matrices. They include O_3^-, SO_2^-, NO_2^- and ClO_2^- and are readily produced by interaction of neutral molecules of moderate or high electron affinity with photoelectrons derived from alkali metal atoms co-deposited in the matrix. The vibration frequencies of the ions usually vary with the alkali metal, showing that a 'partially separated ion pair' is formed.

7.5 Larger molecules

Among larger reactive species of interest that have been trapped in matrices are the perhalo methyl radicals CX_3 and their heavier analogues, CO_3 and some oxides of nitrogen.

Perhalo methyl radicals. Unlike CH_3, which is thought to be planar, all the perhalo methyls are pyramidal. They have been produced or trapped in matrices by a variety of methods, some of which are set out below:

$$CF_2N_2 \xrightarrow{h\nu} CF_2 \xrightarrow{F} CF_3$$

$$CHF_3 \xrightarrow{h\nu} CF_3 + H$$

$$CHCl_3 \xrightarrow{h\nu} CCl_3 + H$$

$$CCl_4 + Li \rightarrow CCl_3 + LiCl.$$

$CBr_4 \xrightarrow{\Delta} CBr_3$ in gas phase followed by trapping

$$CBr_4 + Li \rightarrow CBr_3 + LiBr.$$

Of these methods the lithium atom/halide abstraction is most suspect, in the sense that the inevitable presence of lithium halide in the matrix cage is bound to cause some perturbation of the vibrational spectrum. While photolysis appears to be a simple reaction as written, it is in fact necessary to use such energetic radiation to cause bond cleavage that secondary photolysis and photoelectron production are possible. This results in the formation of other fragments whose spectra must be distinguished from those of the desired species. However, in each case, at least two different techniques give comparable IR data and the radicals can be regarded as reasonably well characterised. The IR data show two stretching vibrations, indicating the radicals are pyramidal.

In the photolysis of chloroform, several charged species (CCl_3^+, $HCCl_2^+$, $HCCl_2^-$) and $HCCl_2$ were identified as well as CCl_3. The formation of these ions probably involves photo-ionisation, as the vacuum UV light used for photolysis has an energy of over 10 eV and could be expected to ionise radical species such as CCl_3 and $HCCl_2$. Negative ions arise from capture of the liberated electrons, or possibly by 'dissociative electron capture':

$$HCCl_3 + e^- \rightarrow [HCCl_3^-] \rightarrow HCCl_2^- + Cl.$$

SiF_3 and $SiCl_3$ are also known in matrices and IR studies show that they are pyramidal. The e.s.r. evidence for non-planarity depends on the magnitude of the ^{29}Si coupling to the unpaired electron spin, which is related to the 's-orbital character' of the orbital containing the lone electron.

Carbon trioxide. CO_3, produced in matrices by reaction of carbon dioxide with mobile oxygen atoms, or trapped among the products of a carbon dioxide discharge, has the interesting structure

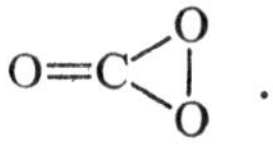

Five of the expected six IR-active fundamentals have been assigned. The C=O frequency occurs near 2000 cm^{-1} while the asymmetric C—O vibration is at 972 cm^{-1}.

Reaction of ground-state oxygen atoms (^3P state) with carbon dioxide to give a singlet (spin-paired) product involves a change of the spin quantum number of the system from 1 to 0 and is thus formally forbidden. However, it seems likely that spin selection rules are less rigorous in the matrix than in the gas phase. Alternatively, the reaction might involve excited singlet-state oxygen atoms, though these might not have been expected to retain their excitation in the matrix long enough for reaction to occur.

Oxides of nitrogen. Oxides of nitrogen stable at room temperature include N_2O, NO, NO_2 and N_2O_5. NO and NO_2 have each a single unpaired electron, but while NO_2 dimerises reversibly to N_2O_4 at moderate temperatures and pressures, NO is effectively monomeric even at lower temperatures. However, in a matrix the dimer N_2O_2, which exists in *cis* and *trans* forms, can be identified by its infrared spectrum. N_2O_3, the 'mixed dimer', is also known in nitrogen matrices, and again isomeric forms are known. The normal

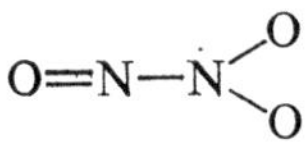

form can be converted by IR photolysis to an 'extended form'

O=N—O—N=O

which is symmetric but not linear. UV radiation reverses this process.

II. HYDRIDE SPECIES

The hydrogen atom itself is uniquely mobile in matrices, which accounts in part for the large number of hydride species observed. Not only may reactive molecules be formed in matrices by loss of hydrogen atoms, singly or in pairs, but the resultant hydrogen atoms readily migrate and often form other products by atom-addition reactions. Matrix materials such as nitrogen and the rare gases are inert to hydrogen atoms but carbon monoxide reacts readily to form HCO, so that a carbon monoxide matrix can be used as a 'hydrogen sink' to ensure that hydrogen atom addition to precursors or product molecules does not occur.

Hydrogen atoms have, of course, no vibrational transitions in the IR or Raman, and the first allowed electronic transition from the ground-state atom occurs at 121.6 nm (Lyman α) in the vacuum UV. They are usually detected only by e.s.r., which shows the expected doublet for ^1H (nuclear spin $I = \frac{1}{2}$) or triplet for ^2D ($I = 1$). The coupling constant A is large (for ^1H 1420 MHz = 0.04 cm^{-1}) but does not vary much from one matrix to another.

7.6 Diatomic hydrides AH

In contrast with gas-phase studies, where electronic spectra of the monohydrides of most atoms have been reported, rather few monohydrides have been studied in matrices, those of most interest being CH, SiH and GeH; NH; OH and SH. The stable hydrogen halides have been extensively studied and show the effects of aggregation, hindered rotation and quantised translation in the matrix.

Group IV monohydrides. CH, SiH and GeH are reported to be formed by photolysis of the hydrides MH_4 in a matrix. CH is also formed by photolysis of CH_3F. The identification of CH is most certain, being based on the observation of three of the known UV bands of the molecule; the 0–0 peaks appear only slightly shifted from the gas-phase values and appear in similar positions on photolysis of CD_4, showing that they do not involve vibration. However, the infrared band expected for the ground-state molecule at 2732 cm^{-1} has not been reported, and photolysis of CH_4 does not appear to lead to e.s.r. signals attributable to CH.

SiH and GeH are comparatively poorly characterised. In each case a single infrared band has been reported; while the H/D isotope shift is consistent with the assignment, the evidence is not conclusive.

Group V monohydrides. NH, with a $^3\Sigma$ ground state (two unpaired π-electrons) has not been observed by e.s.r., though it is formed in high concentrations by photolysis of HN_3 or NH_3. The IR band (at 3132 cm^{-1} in argon) has been observed and the assignment to NH confirmed by ^2D and ^{15}N isotopic substitution. The UV band (corresponding to the $A(^3\Pi) \leftarrow X(^3\Sigma)$ transition) near 340 nm has been studied in detail by several groups; the band moves to lower energies in the matrix, the red-shift increasing in the order argon <

krypton < xenon. The vibration frequency in the excited state also decreases in the same order. By contrast the vibration frequencies in the ground state vary only slightly from the gas-phase value in nitrogen or argon matrices. Clear evidence of essentially unhindered rotation in the matrix is derived from the UV studies; the rotational constant is similar to that observed in gas-phase studies. The fine structure is believed to arise from slow relaxation of rotational excitation: molecules excited by the absorption of radiation will return to the electronic ground state by emission and other relaxation processes, some of which may leave the molecules in rotationally excited state for long enough for absorption by these excited molecules to be observed.

PH and AsH do not appear to have been reported though e.s.r. studies of the photolysis of PH_3 and AsH_3 have been made.

Group VI monohydrides. OH also rotates relatively freely in rare gas matrices; absorption and emission studies of the $A(^2\Sigma^+) \longleftrightarrow X(^2\Pi)$ system in neon show extensive rotational fine structure as well as vibrational structure. Gas-matrix shifts here are again to the red, and again ground-state vibration frequencies are more or less unchanged, whereas the excited state has a lower vibration frequency in the matrix. The OH radical was produced by X-ray bombardment of the matrix. It seems that the UV spectra of OH in other rare gases are so complex as to have so far defied interpretation.

Unfortunately, the IR spectrum of OH in neon has not yet been reported. IR results for other matrix gases seem at variance with the known gas-phase stretching frequency and indeed *two* bands are reported in argon or krypton, shifted by 115 cm^{-1} and 140 cm^{-1} below the gas-phase value. While the double band could be explained on the basis of rotation or rotation/translation quantisation the shift seems larger than expected. It has been suggested that these bands (found after vacuum UV irradiation during deposition) are due to OH^-, but the stretching frequency for this species in ionic solids is even higher than that for gas-phase OH.

The observed bands at 3452 cm^{-1} and 3428 cm^{-1} in argon for ^{16}O, ^{1}H systems gave simple shifts with 1:1 ^{16}O:^{18}O mixtures, showing the presence of only one oxygen atom in the species. Complete substitution of D for H gave a doublet at lower frequencies. All the observed isotopic shifts are compatible with the assignment to OH (or OH^-), but mixed H/D spectra are required to demonstrate that the species contains only one atom of hydrogen.

SH is but poorly characterised in a matrix; it has been reported as a product of photolysis of hydrogen sulphide in argon. The IR frequency is shifted 50 cm^{-1} from the gas-phase value (see above, OH). Some UV transitions were also claimed, but few details are available.

7.7 Triatomic dihydrides AH_2

Again relatively few simple dihydride reactive species have been reported, the best attested being SiH_2, GeH_2, NH_2 and PH_2. The most notable absentee is of course CH_2, which is well characterised in the gas phase but which has been searched for without success in matrices.

Group IV dihydrides. CH_2 is commonly prepared as a transient intermediate by photolysis of diazomethane, CH_2N_2, and can also be produced by photolysis of ketene, CH_2CO, or of methane. The first two of these precursors decompose on photolysis to give CH_2 and N_2 or CO. It appears that in a matrix photolysis gives initial decomposition which is reversed quantitatively because the by-product cannot escape from the matrix cage. Indeed, if CH_2N_2 is photolysed in a carbon monoxide matrix CH_2CO is produced, while if CH_2CO is photolysed in a nitrogen matrix CH_2N_2 is produced. Decomposition of methane *in situ* should give a better method of production since the hydrogen atoms are free to migrate. Photolysis of methane is only possible using very energetic (vacuum UV) radiation; while CH_2 is probably formed along with CH, CH_3 and H, it appears to decompose photolytically or react with migrating hydrogen atoms so readily that it cannot be detected. Photolysis of methane in nitrogen, however, gives CH_2N_2, implying that CH_2 is at least formed transiently.

SiH_2 and GeH_2 seem to be less reactive than CH_2; this may be associated with their singlet electronic ground states (CH_2 has a triplet ground state). They are formed during the UV photolysis of SiH_4 or GeH_4 and have been identified by IR isotopic studies in rare gas matrices.

Group V dihydrides. NH_2 is one of the classic reactive species in matrix studies. It was one of the first species to be reported trapped in rare gas matrices (in 1958), it has been studied by IR, UV/visible and e.s.r. spectroscopy, and has been prepared by

(i) discharge before trapping ($N_2H_4 \rightarrow NH_2$),
(ii) photolysis in a matrix ($NH_3 \rightarrow NH_2 + H$),

(iii) reaction after photolysis ($HN_3 \rightarrow NH+N_2$; $NH+H \rightarrow NH_2$). In addition to the electronic, vibrational and electron spin resonance transitions that have been observed, the rotation behaviour has been inferred from fine-structure in the electronic spectrum and the details of the e.s.r. spectrum. Rotation occurs about an axis parallel to the H–H vector.

PH_2 has been prepared, together with PH_4, by photolysis of phosphine. It too rotates in krypton matrices, as shown by the e.s.r. spectrum. IR and electronic spectra have also been reported. AsH_2 was not observed when arsine was photolysed under the same conditions.

7.8 Trihydrides AH_3

Only the group IV derivatives CH_3, SiH_3, GeH_3 and SnH_3 have been reported, though among stable molecules NH_3 is interesting as studies show that rotation can occur in argon but not in nitrogen matrices at quite low temperatures. The group III trihydrides (BH_3, AlH_3...) do not seem to have been prepared in matrices, and H_3O remains elusive.

Group IV trihydride radicals. CH_3, SiH_3, GeH_3 and SnH_3 can be prepared by γ-irradiation of the parent hydrides MH_4 in a matrix; the e.s.r. spectra are interpreted in terms of a planar structure for CH_3 and pyramidal structures for SiH_3 and GeH_3. CH_3, SiH_3 and GeH_3 are also formed by vacuum UV photolysis of the hydrides MH_4, and have been studied by IR methods; the 150 nm band of CH_3 was also observed in the UV. These studies are said to be consistent with a planar structure for CH_3 and pyramidal structures for SiH_3 and GeH_3.

It is interesting to note that the symmetric deformation frequency of CH_3 (see fig. 7.1) is considerably *smaller* than that of SiH_3 or GeH_3. An abnormally low bending frequency is often associated not with perfect planarity, but with the existence of a nearly planar structure with a low barrier to inversion. Comparison of the bending frequencies with those of the well-characterised NH_3 and PH_3 species (see table 7.2) suggests that CH_3 could possibly be strictly non-planar with a low barrier to inversion.

The first reports claiming to have prepared CH_3 in a matrix involved the reaction of CH_3X ($X = I$, Br, Cl) with co-deposited lithium atoms. At the time the results appeared reasonable, but the deformation frequency claimed at 730 cm^{-1} was much higher than

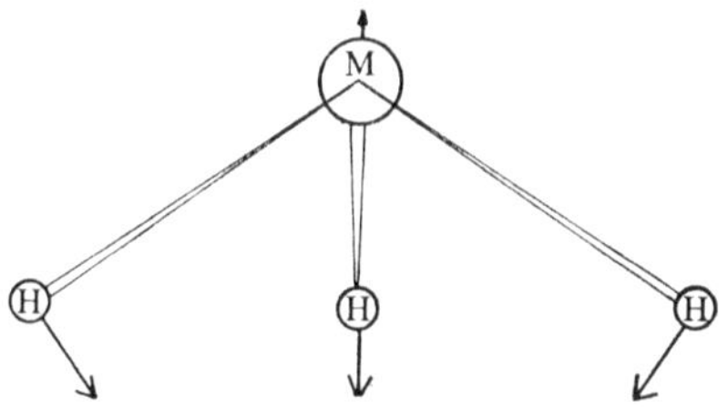

Fig. 7.1. Symmetric deformation of MH_3.

TABLE 7.2. *Symmetric deformation frequencies of trihydride species*

MH_3	ν_2/cm^{-1}	Barrier to inversion	Structure
SiH_3	950	High	Pyramidal
GeH_3	850	High	Pyramidal
CH_3	615	Low?	Near planar?
PH_3	990	High	Pyramidal
NH_3^+	990		Planar
NH_3	950	Low	Pyramidal
PH_3^+	480	Low	Near planar

that found when CH_3 was produced by photolysis of methane. Further work showed that the band shifted when different alkali metals were used, and it is now assigned to a species CH_3MX (M = Li, Na, K). 'Atom-abstraction' reactions are thus not necessarily a reliable means of preparing matrix-isolated reactive species. CH_3Li has also been identified among the products of the reaction of methyl iodide or bromide with lithium in a matrix; the ^{12}C–^{6}Li stretching frequency is 558 cm^{-1}.

7.9 Triatomic monohydrides HAB

Corresponding to the 16-electron rule for triatomic molecules ABC, the '10-electron rule' states that a molecule HAB with 10 or fewer valence electrons will be linear, whereas with more than 10 it will be bent. This is because of the necessity for the π^*-antibonding orbital to be occupied if the molecule were linear. It may also be inferred that addition of atomic H to diatomic AB will strengthen the AB bond if HAB contains 10 valence electrons or less, and weaken it if HAB contains more than 10 valence electrons.

Both predictions are borne out by the experimental evidence, which is set out in table 7.3. In practice, the AB stretching frequency is

TABLE 7.3. *Characteristics of some triatomic hydrides: HAB*

AB	HAB	No. of electrons in HAB	Linear/ bent	νAB	ν(H)AB	νHA(B)
CC	HCC	9	L	1840	1850	
CN	HCN	10	L	2046	2093	3303
CN	HNC	10	L	2046	2030	3620
CO	HCO	11	B	2160	1860	2482
CF	HCF	12	B	1280	1180	2918
NO	HNO	12	B	1890	1560	3590
NF	HNF	13	B	1115	1000	
OO	HOO	13	B	1580	1100	3411
OF	HOF	14	B	1030	884	3483

not much altered by addition of H if HAB contains 10 or fewer electrons; this is partly due to coupling with the HA vibration which tends to reduce the frequency of the AB vibration. However, if HAB contains more than 10 electrons, the AB frequency is markedly lowered.

All the triatomic monohydrides listed except HCN are short-lived species; HOF can be handled as a vapour with care, but the remainder can only be studied in matrices or flash systems. A notable absentee is HN_2, with 11 electrons; no evidence for its formation in a matrix (or in the gas phase) seems to exist. The boron-containing species HB_2, HBC, HBN, HBO and HBF would also be interesting; all but HBF should be linear.

Most of the species listed in table 7.3 have been identified in matrices by IR spectroscopy, isotopic substitution confirming the assignments. In some cases electronic spectra are also observed. E.s.r. spectra have been reported for HC_2, HO_2 and HCO, each of which has a single unpaired electron. HO_2 is said to be free to rotate in argon at 4 K.

Some species containing heavier atoms, such as HNSi and HCCl, analogous to species in table 7.3 are known from matrix studies. The formation of HCCl involved insertion of migrating carbon atoms, produced by double photolysis of N_3CN, into HCl in argon or nitrogen:

$$N_3CN \xrightarrow{h\nu} N_2 + NCN \xrightarrow{h\nu} C + N_2$$

$$HCl + C \rightarrow HCCl.$$

HCCl, like HCF but unlike CH_2, has a singlet ground state.

Two other interesting species have been reported to be formed from hydrogen halides in matrices. HCl is reported to give ClHCl, with a linear symmetric structure, when a mixture of HCl, chlorine and argon is discharged before condensation. When HBr and bromine are used a similar species, said to be BrHBr is formed; this has the same IR spectrum as the product of photolysis of trapped HBr, which is said to be the ion $BrHBr^-$. It has been suggested that the spectrum assigned to ClHCl is also due to the negative ion. A mechanism for the formation of the ions involves photoelectron capture by HX dimer, with loss of hydrogen atoms:

$$(HX)_2 + e^- \rightarrow XHX^- + H.$$

Two different explanations are thus suggested for the same experimental observations; it is not easy to see how this problem could be resolved.

7.10 Higher hydrides H_nAB ($n > 1$)

Species of interest in this category include N_2H_2, H_2CN and H_2CNH.

Diimide, HN=NH, exists as *cis* and *trans* forms in matrices when produced by photolysis of HN_3. As diffusion appears not to be necessary for its formation, one may deduce that photolysis of dimers of HN_3 is involved:

$$(HN_3)_2 \xrightarrow{h\nu} HNNH + 2N_2.$$

The Raman spectrum of the *trans* isomer has also been observed in a nitrogen matrix and the vibrational spectra assigned on the basis of isotopic substitution. The N–N double bond stretching frequency is at about 1530 cm^{-1}. It is possible that the molecule has a triplet ground state, though no e.s.r. results have been reported.

The e.s.r. spectrum of the methylene imino radical, H_2CN, has been reported. It is formed when hydrogen atoms, produced by photolysis of hydrogen iodide in the argon matrix, react with HCN:

$$H + HCN \rightarrow H_2CN.$$

Methylene imine itself, H_2CNH, may be produced by the photolysis of methyl azide in a matrix:

$$CH_3NNN \xrightarrow{h\nu} [CH_3{-}N:] \rightarrow CH_2{=}NH.$$

An alternative method involves the photolysis of diazomethane. It is thought that CH_2 is produced (see above) which reacts with a neighbouring CH_2N_2 to give H_2CNH and HCN:

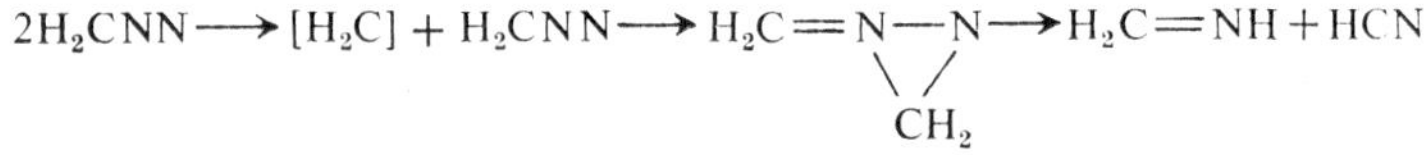

7.11 Hydrides of formula HABC

Nitrous acid itself, formulated as HONO, does not contain much of this species in its molecular form. HONO has, however, been studied in matrices; it may be formed by reaction of hydrogen atoms with NO_2 or by photolysis of HN_3 in the presence of oxygen in nitrogen matrices. The *cis–trans* isomerism has been studied in detail by IR methods.

Just as photolysis of HCN leads to the isomer HNC, so photolysis of HNCO gives the cyanic acid isomer HOCN. Its vibrational spectrum has been studied in argon. It appears that photolysis initially gives NH+CO, not H+NCO, as no absorptions due to NCO appear, although hydrogen is expected to be mobile under the conditions used. The formation of a cyclic intermediate addition product

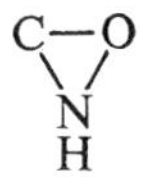

is postulated followed by rearrangement to HOCN or HNCO.

A similar reaction of OH, produced by photolysis of water in a carbon monoxide matrix, is said to give HOCO, which exists in *cis* and *trans* forms. Both forms can be photolysed to give CO_2.

8 Unusual metal compounds

In this chapter we shall discuss some unusual molecules which illustrate novel aspects of the chemistry of metals. In the first section we examine some of the small molecules which would normally be found in polymeric form. In the latter part of the chapter we are concerned with unusual transition metal complexes.

I. MONOMERS AND SMALL OLIGOMERS OF NORMALLY POLYMERIC MATERIALS

8.1 Metal halides

Alkali metal halides MX and oligomers. Alkali metal halides can be matrix isolated, either from the vapour following evaporation or by means of chemical reactions in the matrix. The best studied are the lithium halides; little work has been published on sodium halides and the heavier metal compounds.

Lithium fluoride has been studied in great detail by IR methods. Evaporation followed by trapping gives spectra with several bands, which have been analysed in terms of monomer, dimer and trimer species. The dimer and trimer are thought to be small rings, though an alternative linear structure for the dimer has been suggested. The other lithium halides form similar oligomers, presumably during deposition, as they appear to be present even when the vapour is super-heated.

The vibration frequencies found for the monomers are quite markedly sensitive to the matrix, and are all reduced considerably from the gas-phase value. These large shifts have been attributed to electrostatic effects, as they depend on the polarisability of the matrix material.

As the force between a dipole and a polarisable atom depends on the distance between them as well as on the magnitudes of the dipole and the polarisability (α), no simple relationship between the shift and α is expected. The shift does, however, increase fairly smoothly

TABLE 8.1. *Polarisabilities of gases and matrix shifts for ^7LiF*

	$\alpha/10^{-24}$ cm^3	Shift $(\nu_m - \nu_g)$/cm^{-1}
Ne	0.4	−43
Ar	1.6	−67
Kr	2.5	−77
Xe	4.0	−87
N_2	1.7	−132
CH_4	2.6	−126

TABLE 8.2. *Dipole moments of HX and LiX/debye*

X	HX	LiX
F	1.9	6.6
Cl	1.1	8.3
Br	0.8	6.2
I	0.4	7.1

down the group neon, argon, krypton, xenon (see table 8.1). Molecular matrix materials such as nitrogen and methane give rise to much greater shifts than do the rare gases of comparable polarisability (argon and krypton respectively).

It is interesting to note that the shifts for all the lithium halides seem to be approximately the same for a given matrix; while the dipole moments of the lithium halide monomers are all about the same (table 8.2), their bond lengths differ, so again no such simple relationship would be expected. It is worth noting though that the hydrogen halides, with much lower dipole moments, give much smaller shifts in matrices than do the lithium halides.

Dihalides MX_2. The structure of a symmetrical dihalide MX_2 may be linear or bent; if it is linear the symmetric stretch ν_1 is inactive in the IR, whereas if it is bent all three vibrations should be observed in the IR.

It is found that for BeF_2 only one stretch (ν_3 at 1530 cm^{-1}) and the bend (ν_2 at 325 cm^{-1}) are observed in the IR, which is consistent with molecular beam deflection evidence showing that BeF_2 is linear. On the other hand MgF_2, CaF_2, SrF_2 and BaF_2 are said to show two stretching bands in the IR, showing that they are bent. The vibration frequencies and bond angles, derived from a normal coordinate analysis, are given in table 8.3.

TABLE 8.3. *Vibration frequencies and calculated bond angles for* MF_2

M	ν_1/cm^{-1}	ν_2/cm^{-1}	ν_3/cm^{-1}	Angle/°	$(\nu_3-\nu_1)$/cm^{-1}
Be		325	1530	180	?
^{24}Mg	478	242	837	158 ± 2	359
^{40}Ca	485	163	554	140 ± 5	69
^{86}Sr	441	82	443	108 ± 5	2
Ba	413		390		−23

It is striking that the difference in the stretching frequencies $\nu_3-\nu_1$, decreases rapidly down the group and is probably negative for BaF_2 (ν_3 is the strongest band for all the fluorides). (See note p. 137).

8.2 Covalent halides

Many covalent halides (particularly fluorides) are known to be highly associated in the solid and liquid states, and even in the vapour, so the analysis of their vibrational spectra is not easy. Often no firm conclusions about the structures of isolated molecules can be drawn, simply because no isolated molecules exist in conventional samples. Matrix isolation can in principle be used to overcome this; if a matrix is deposited from vapour containing predominantly monomeric species it is reasonable to hope that the vibrational spectrum found will be largely that of the monomer.

IR studies on these lines have been made of ClF_3, BrF_3, SF_4, BrF_5, AsF_5 and SbF_5. In all cases but the last, isolation of monomers occurs and the selection rules and vibration frequencies are consistent with the predictions of simple electron-pair repulsion theory. Dimers were detected for ClF_3, BrF_3, SF_4 and BrF_5 and their spectra interpreted in terms of fluorine bridges which may well be involved in fluorine exchange, known to occur in the liquids. No association was observed for AsF_5.

For SbF_5 a very different spectrum was obtained which was initially assigned to a C_{4v} (square pyramidal) monomer. It has since been pointed out, however, that the spectrum reported is closely similar to that of the vapour at room temperature, which contains no monomer, so it must be reassigned as due to an exceptionally stable dimer or higher oligomer. The structure of *monomeric* SbF_5 is almost certainly D_{3h} (trigonal bipyramidal).

IR studies of $NbCl_5$ in cyclohexane and nitrogen matrices showed that depending on the conditions of formation either monomer or dimer could be isolated. Specifically, deposition from the vapour gave monomer spectra with matrix ratios of about 100:1 or more, while with smaller matrix ratios dimers predominated. Monomer spectra were consistent with D_{3h} symmetry, while the spectrum of the dimer (which persists in solution and in the crystalline solid) was interpreted in terms of 6-coordinate niobium with two bridging chlorine atoms.

8.3 Oxides of metallic elements

Alkali metals. The reaction of oxygen with the alkali metals under normal laboratory conditions gives one or more of a number of solid products: M_2O, M_2O_2, MO_2 and other less well-characterised compounds. The tendency to form M_2O_2 and MO_2 increases as the size of the alkali metal increases. Thus LiO_2 has not been prepared at room temperature, and Cs_2O is not found among the products of reaction of caesium with oxygen or air under normal conditions.

By contrast, all the alkali metals (evaporated from furnaces) react with molecular oxygen in inert matrices to form molecular species containing pairs of oxygen atoms such as MO_2, M_2O_2 and MO_4. Lithium gives only LiO_2, sodium NaO_2 and Na_2O_2, while potassium, rubidium and caesium give MO_2, M_2O_2 and MO_4.

The molecular monoxides M_2O can be matrix isolated from the vapour over heated solids of this formula; the matrix-isolated samples also show bands due to MO_2, MO and M_2O_2, but which, if any, of these is present in the vapour is unclear. M_2O is also formed by reaction of M atoms with N_2O in a matrix; elimination of N_2 leaves MO or M_2O.

Vibrational studies have thrown considerable light on the structures of the various species in the M/O systems (see fig. 8.1) and it is worth examining the results in more detail.

Molecular MO

The oxides MO are interesting in that they are similar to the fluorides, containing one fewer electron. The stretching frequencies are given in table 8.4.

M_2O

The structures deduced for M_2O in matrices vary; Li_2O is found to be nearly linear, while the bond angles for K_2O, Rb_2O and Cs_2O are calculated to be 158°, 156° and 125° respectively. These precise figures are subject to considerable uncertainty, due to the enforced neglect of anharmonicity in the calculations.

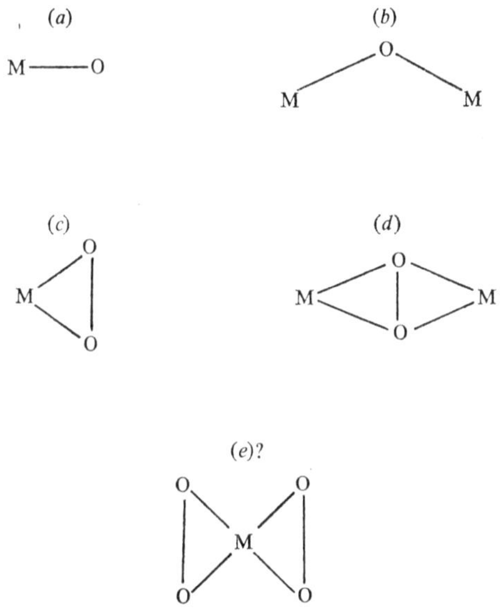

Fig.8.1. Alkali metal (M) oxide species observed in matrices.

TABLE 8.4. *Stretching frequencies of MO molecules*

	ν/cm^{-1}	Matrix
^{7}LiO	740	N_2
KO	384	N_2
CsO	314	N_2

MO_2

The symmetric triangular superoxides MO_2 are held to be largely ionic ($M^+O_2^-$) as the O–O stretch is very weak in the IR and strong in the Raman, while the M–O stretching modes are observed more strongly in the IR than in Raman. The O–O stretch is not greatly affected by changing the alkali metal (see table 8.5).

E.s.r. spectra are also consistent with an ionic formulation for NaO_2 in a matrix.

M_2O_2

The molecular peroxides form minor constituents of the matrix-isolated sample in most experiments, and their spectra have not been convincingly assigned. Two distinct sets of three IR bands have been

TABLE 8.5. *Vibration frequencies (/cm^{-1}) for MO_2*

	ν OO	ν_s MO	ν_{as} MO	$\nu_s - \nu_{as}$
^{7}Li	1097	699	492	207
Na	1094	391	333	59
K	1108	307		0?
Rb	1110	255	282	−27
Cs	1114	236	269	−33

TABLE 8.6. *Frequencies (/cm^{-1}) and structures for Li_2O_2*

	Ohio results	Virginia results
ν_4	242	298
ν_5	325	796
ν_6	522	446
OLiO angle	116°	55°
Structure	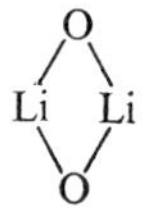	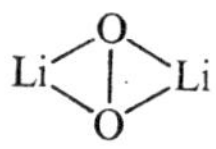

assigned to the three IR-active fundamentals ν_4, ν_5 and ν_6 of Li_2O_2 by two different groups; these assignments (see table 8.6) lead to essentially opposite conclusions about the geometry of the molecule. The 'Virginia' structure seems more chemically reasonable, as Li_2O_2 is formed by reaction of a second atom of lithium with LiO_2 in the matrix. The observation of a Raman spectrum attributable to this species might help to resolve the problem.

MO_4

A number of structures are possible for this molecule. It is clear, from isotopic studies, that the two pairs of oxygen atoms retain their individuality, and that the two O atoms in each pair are equivalent. Further evidence is required to establish the structure.

Another species, also involving two pairs of oxygen atoms, is $(MO_2)_2$, a weakly interacting dimer of MO_2 molecules. The main effect of the second molecule appears to be an enhancement of the IR activity of the O_2^- stretch.

Transition metals. The study of the vapours over heated transition metal oxides or of the products of reaction of gaseous oxygen with heated transition metals shows that a variety of simple species can be isolated in matrices. Unfortunately, the electronic structures of small molecules of this type, often containing partially filled d-subshells, are complex and the resulting spectra are not easy to interpret. However, considerable progress is being made, and spectra of a number of diatomic species MO and a few dioxides MO_2 have been reported. We shall consider only a few of these here.

Tungsten oxides

Tungsten, heated to 2000 K in a stream of oxygen, or any oxide of tungsten heated above 1600 K, gives vapours containing WO, WO_2 and WO_3, which can be trapped in a matrix by co-condensation with neon or argon. Polymeric species such as W_2O_6, W_3O_8, W_3O_9... were also found. The proportions of the various species depended, as might be expected, on the conditions of production.

WO was reported to have a vibration frequency of 1055 cm^{-1}. This seems extremely high (cf. CsO: $\nu = 314\ cm^{-1}$ in nitrogen), and if correct implies a very strong bond (a triple bond is possible electronically).

Emission and absorption spectra of WO in the visible and ultraviolet regions were also observed; they were correlated with the more complex spectra observed in vapour-phase studies so WO is well characterised in the matrix. The same cannot be said of WO_2 or WO_3, where no gas-phase spectra are known. Although the electronic transitions attributed to them could arise from such molecules it is by no means proved that they do. Unfortunately, insufficient isotopic mixing experiments were carried out to confirm the formulae. Infrared bands were also attributed to WO_2 and WO_3 on the basis of complete ^{18}O substitution only, so the assignments are tentative.

Tantalum oxides

A similar situation exists for the tantalum oxides. It is thought, from mass-spectrometric studies in the vapour phase, that TaO and TaO_2 are both present in the vapour over heated Ta_2O_5 or tantalum in oxygen, and the spectra of TaO in matrices compare well with the gas-phase spectra. This is especially so for neon matrices, where very narrow bands are observed. These studies show that the ground state of TaO is $^2\Delta$, with a single electron in the tantalum 5d-level. A high stretching frequency is again found in the IR, confirming the value determined from the gas-phase emission spectrum, 1020 cm^{-1}.

Two electronic bands attributed to TaO_2 are observed: they consist of progressions in a vibration frequency of 280 cm^{-1}, presumably a bend, and are thus due to a molecule with at least three atoms. One also has a weaker progression in the same frequency separated from the main one by 935 cm^{-1}, suggesting a Ta–O stretching vibration. Two bands in the IR, at 971 cm^{-1} and 912 cm^{-1}, shift as expected for TaO_2 on complete substitution of ^{18}O, but the pattern with partial substitution is more complex than expected, and it seems likely that two non-equivalent oxygen atoms are present. The species TaO_2 cannot be said, then, to be adequately characterised.

Uranium oxides

Oxides of uranium have also been matrix isolated recently. Vaporisation of UO_2 and co-condensation with matrix material gave spectra attributed to UO and UO_2 molecules in the IR. In this case partial ^{18}O substitution confirms that a band at 874 cm^{-1} could be due to ν_3 of UO_2, as it gives a triplet; ν_2, the bend, is assigned to a band at 81 cm^{-1}, but ν_1 was not observed. The bond angle was estimated to be about 108°. UO is assigned a stretching frequency of 776 cm^{-1}, which is lower than ν_3 of UO_2 (contrast the assignments for TaO and TaO_2).

8.4 Oxides of less-metallic elements

Oxides of group III metals – aluminium, gallium, indium, thallium. This group of oxides exemplifies to a striking degree the possibility that vapour species may be quite different from the solids they arise from. The stable oxides in the solid phase are M_2O_3, though Tl_2O can be prepared in the absence of air, but the vapours formed by reaction of metal with oxygen or by heating mixtures of metal and M_2O_3 contain almost entirely M_2O. The M_2O molecules have been isolated and studied in detail. Boron, incidentally, shows different behaviour, in that B_2O_3, B_2O_2 and BO_2 can all be identified in matrices, while B_2O is not observed.

Al_2O is produced, then, as a vapour when aluminium is heated with Al_2O_3. It has been isolated in rare gas matrices and gives a strong band at 994 cm^{-1} in argon, shifting to 951 cm^{-1} on ^{18}O substitution. This is assigned to ν_3, the asymmetric stretch, expected to be stronger and at a higher frequency than ν_1. The latter is assigned to a weaker band at 715 cm^{-1} in argon, shifting to 700 cm^{-1} on ^{18}O substitution. No other bands were observed to arise on partial substitution, showing that only a single oxygen atom is present in the molecule. The bending mode ν_2 has not been observed, but the bond angle is calculated as 145° from the ^{18}O shift of ν_3. This represents a *lower* limit, and it is possible that Al_2O is linear as predicted from Walsh's treatment. In this case ν_1 is IR-forbidden and the band at 715 cm^{-1} must be assigned to some other species.

Bands assigned to ν_3 for Ga_2O, In_2O and Tl_2O have also been observed in nitrogen matrices; the species were formed by high temperature reaction of M with oxygen or M with M_2O_3 and co-condensed with the matrix. The $^{16}O/^{18}O$ shifts gave estimated bond angles of 143°, 135° and 131° respectively, again representing lower limits. Soon after, another report gave infrared values for all three vibrations of Al_2O, Ga_2O, In_2O and Tl_2O, which were said, on the basis of the positions of ν_1 and ν_2, to imply models with very small MOM angles, and substantial metal–metal bonding.

Further work by the earlier authors has now shown this assignment to be erroneous; $^{16}O/^{18}O$ mixtures lead to *triplets* for the bands assigned to ν_1 and ν_2 of Ga_2O and In_2O, showing that each is due to a species containing *two* oxygen atoms. It seems likely that this is a dimer M_4O_2, perhaps containing a four-membered ring

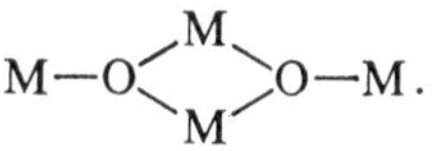

The two bands concerned in each spectrum vary together in intensity and in particular *increase* in intensity relative to ν_3 as diffusion becomes more important.

The bending modes ν_2 are not observed and probably lie below 200 cm^{-1}, the limit of the region studied. Several very weak bands remain as candidates for ν_1; unfortunately their weakness would make it very expensive to observe $^{16}O/^{18}O$ shifts.

In the interesting mixed molecule GaOIn, produced by heating mixed oxides, again only one stretching vibration is observed, falling between ν_3 of Ga_2O and ν_3 of In_2O. The presence of only *one* mixed monomer confirms the species as M_2O.

Mixed gallium/indium studies on the dimer indicate a total of *seven* mixed molecules, which is consistent with the proposed structure for M_4O_2.

The thallium oxide system is more confused, as three sets of contradictory data have been presented. One group report ν_3 of Tl_2O in nitrogen at 625 cm^{-1}, and an ^{18}O shift consistent with a bond angle of 131° (again, a lower limit). A strong band near 500 cm^{-1} was shown to be due to a species containing two oxygen atoms, as a triplet appeared when mixed $^{16}O/^{18}O$ systems were studied. Another group took this band as ν_1 and also reported ν_3 at 643 cm^{-1} and ν_2 at 383 cm^{-1} in argon, again leading to a very narrow bond angle and metal–metal bonding. A third study assigned bands at 643 cm^{-1} and 571 cm^{-1} to ν_3 and ν_1, and calculated the bond angle as 90° and ν_2 near 130 cm^{-1}.

There is thus general agreement over ν_3, though the shift between argon and nitrogen is larger than usual, but complete divergence over the bond angle and the bending frequency. The assignment to ν_1 and ν_2 by the second group has been shown to be incorrect, but the differences between the first and third group remains to be resolved.

Group IV – silicon, germanium, tin, lead. The main oxides of these elements are MO_2, which are macromolecular solids. Other oxides such as MO also form strongly bonded solids. In the gas phase SiO_2 and all MO have been identified. Matrix isolation studies have considerably extended our knowledge of the oxides of group IV.

Heating SiO_2 gives SiO as the main volatile species; mixtures of silicon and SiO_2 can be evaporated essentially completely as SiO at ~ 1600 K. The vapours have been co-condensed in matrices and the vibrational spectra studied in detail with wide variation in conditions of evaporation and condensation, diffusion and isotopic substitution. Silicon, with naturally occurring isotopes ^{29}Si and ^{30}Si (~ 5 % each) in addition to the major ^{28}Si, is well suited to isotope studies; in this case a band at 1224 cm^{-1} in nitrogen had satellites as expected for silicon isotopic effects and shifted as expected on ^{18}O substitution. It was assigned to SiO. On diffusion this band decreased, showing that it was due to a mobile (i.e. small) and reactive species. Bands at 805 and 766 cm^{-1} varied in intensity together as conditions changed, decreasing on diffusion. They each gave rise to a triplet on partial ^{18}O substitution and were assigned to Si_2O_2. Weak bands at 973, 631 and 312 cm^{-1} were assigned to Si_3O_3 on the basis of ^{18}O substitution; they *increased* in intensity on diffusion.

Force-constant calculations suggested that Si_2O_2 has a planar ring with D_{2h} symmetry; two IR-active Si–O stretches are then anticipated, as observed. Si_3O_3 is also thought to have a planar ring (see fig. 8.2).

Similar molecules were found in matrices prepared from GeO_2, SnO_2 and PbO_2 or by reaction of the metals with oxygen. In addition bands attributed to M_4O_4 were observed, and a T_d structure of two interlocking tetrahedra suggested for Pb_4O_4.

Co-condensation of tin atoms and oxygen in matrices gave spectra assigned to the new isolated molecule SnO_2. The assignment, which

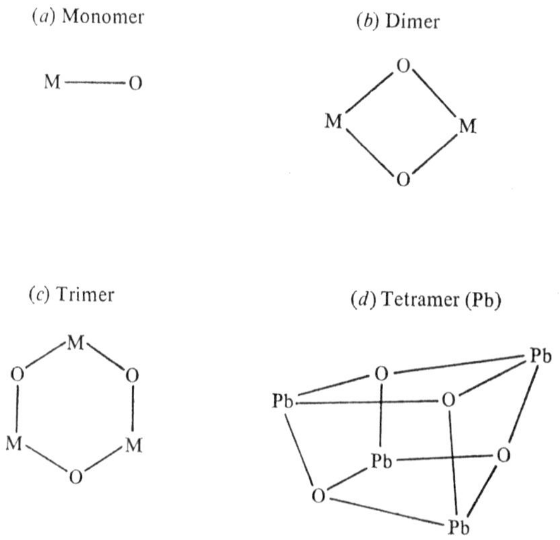

Fig. 8.2. Group IV oxide species.

implied $D_{\infty h}$ symmetry as for CO_2, was confirmed by ^{18}O substitution and the complex tin isotope pattern. Reaction of SnO_2 with tin in the matrix is said to give Sn_2O_2, which has also been observed in Mössbauer studies.

Groups V and VI – arsenic, antimony, bismuth; selenium, tellurium. While the group IV oxides grow more volatile down the group, those of groups V and VI grow less volatile. Thus As_4O_6 and SeO_2 are readily volatile solids that have been studied extensively in the gas phase, while the heavier elements give less well-characterised vapours. Matrix isolation methods have been applied to the study of all these systems.

The IR and Raman spectra of As_4O_6 isolated in nitrogen matrices show that it is present as tetrahedral molecules, as found in the gas phase. The frequencies are closely similar to those found in the vapour but are noticeably different from those of the solid, where the structure is polymeric. The spectra of matrices prepared from the vapours over Sb and Bi oxides have also been studied, but are more complex than those arising from arsenic oxide. The nature of the species present is not yet clear.

Selenium/oxygen system

Crystalline SeO_2 has a chain structure 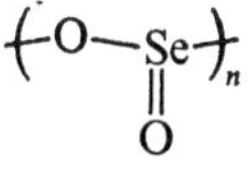 but the vapour contains bent triatomic molecules with a bond angle of 113° 50′, determined very accurately from the microwave rotation spectrum. In matrices, IR and Raman spectra show that isolated molecules are present. The isotopes of selenium (76, 77, 78, 80 and 82) give a characteristic splitting that can be used to derive a *lower limit* to the bond angle. On the other hand ^{18}O substitution can give an *upper limit* to the bond angle. The results from the two isotopic studies give angles of $109° \pm 4°$ and $117° \pm 2°$ for the lower and upper limits respectively. The average of 113° would seem to represent little perturbation from the gas phase. ^{18}O substitution of SeO_2 was achieved by passing oxygen through a microwave discharge (giving O atoms and excited molecular O_2) and immediately over selenium.

Mutual exclusion in the IR and Raman would seem to indicate that a dimer species $(SeO_2)_2$ was centrosymmetric. The structure of the dimer could be based on a four-membered ring, but this is not certain.

Tellurium/oxygen system

TeO_2 only evaporates near 1000 K and the vapour contains various species. Spectra produced by matrix-isolated samples of 'TeO_2 evaporate' are thus more complex than those found for SeO_2. TeO, TeO_2 and Te_2O_2 have been identified in matrices, and polymeric species are also present. TeO_2 has been studied in detail, all three vibrations having been observed; the bond angle is around 109°. The Te_2O_2 species appears to be cyclic (cf. Si_2O_2 above).

In all such studies one should remember that the species identified in the matrix may not accurately reflect the composition of the vapour because polymerisation can occur on condensation, and because only small fragments can be conclusively identified, in most cases, by vibrational spectroscopy.

II. TRANSITION METAL COMPLEXES AND SIMILAR SPECIES

Two main methods have been used to prepare matrices containing novel transition metal complexes and related compounds. They have been developed side-by-side, beginning about 1970, and each has some features that are especially interesting. The first we shall consider is that of photolysis of trapped stable precursors, which has been used so successfully for main group compounds. The second involves the use of reactive matrices either with photolysis of trapped stable precursors or with co-condensation of metal atoms.

8.5 Photolysis of transition metal carbonyls

At first sight the simple carbonyls offer an attractive set of compounds for photolysis studies. All absorb in the quartz UV ($\lambda > 200$ nm), some even in the visible ($\lambda > 400$ nm) and the strength of the absorption bands means that photolysis studies using weak almost monochromatic light sources are possible. On the other hand, they have characteristic strong 'carbonyl stretching' bands in the IR between 2100 cm^{-1} and 1800 cm^{-1} which offer ideal means of identifying the species formed, as much is now known about the characteristic patterns of bands associated with various arrays of carbonyl ligands.

It is perhaps surprising, then, to find that almost as much confusion and controversy has arisen in this field as in others, with different groups of workers assigning their spectra to different compounds, or disagreeing as to the effects of isotopic substitution.

The best known mononuclear carbonyls are $Ni(CO)_4$, $Fe(CO)_5$ and $M(CO)_6$ where M = Cr, Mo or W. The main results are that a single carbonyl is lost when any of these compounds is photolysed with $\lambda > 200$ nm, producing $Ni(CO)_3$, $Fe(CO)_4$ or $M(CO)_5$. An interesting feature is that production of $Fe(CO)_4$ or $M(CO)_5$ is *reversed* by irradiation at longer wavelengths. This may be due to electronic excitation of the fragment to a state whose recombination with neighbouring CO is no longer symmetry forbidden (as it is in the ground state). An alternative explanation involves 'local heating' of the matrix following absorption and relaxation by the fragment permitting the close approach of CO and subsequent recombination.

The structures of the fragments are of interest:

Nickel carbonyls

$Ni(CO)_4$ has a tetrahedral array of carbonyl groups, and $Ni(CO)_3$ appeared to be non-planar with C_{3v} symmetry. This conclusion was based on the observation of two IR carbonyl bands, at 2065 cm^{-1} (weak) and 2016 cm^{-1} (strong) assigned respectively to the A_1 and E modes. About 50 % recombination occurs if the argon matrix is allowed to warm to 30 K. The intensity of the band due to free CO confirms that only a single CO molecule is lost.

Iron carbonyls

$Fe(CO)_4$, together with free carbon monoxide, is produced on photolysis of $Fe(CO)_5$ for a few minutes. The process is reversed by irradiation even with the IR source! The structure is said to be of C_{3v}

symmetry, with a unique 'axial' carbonyl. Prolonged photolysis causes the bands attributed to $Fe(CO)_4$ and free CO to shift slightly, and the photolytic reaction is no longer easily reversible. This suggests that some 'escape' of CO from the immediate vicinity of the fragment occurs in a small proportion of incidents.

Photolysis of $Fe_2(CO)_9$ appears also to lead to loss of carbonyl; the product $Fe_2(CO)_8$ shows both terminal and bridging carbonyl bands. $Fe_2(CO)_8$ (like $Co_2(CO)_8$) may exist in two forms, one with two bridging carbonyls and the other with none.

Chromium, molybdenum and tungsten carbonyls

The results for $M(CO)_5$ favour in general the C_{4v} (pyramidal) structure, the suggestion of an alternative D_{3h} (trigonal bipyramidal) structure being apparently based on impurity or polymer bands. The number of bands observed (three) and the shifts for ^{13}C and ^{18}O substitution are consistent with the C_{4v} form.

Some work has been done on the photolysis of substituted carbonyls such as $CH_3Mn(CO)_5$ and $PR_3Mo(CO)_5$. These systems, too, seem to lead to carbonyl loss rather than cleavage of the bond to the unique ligand. It will be interesting to see further developments in this important area.

Carbonyl anions

A very recent development involves the production of carbonyl anions by photo-ionisation coupled with photolysis. This process occurs to some extent if light with $\lambda < 200$ nm is used for photolysis, leading presumably directly to $M(CO)_{n-1}^{-} + CO^{+}$. Higher yields are obtained if co-deposited sodium or potassium atoms are present. In this case photoelectron transfer can occur with 'normal' UV light ($\lambda > 200$ nm) by the processes:

$$M(CO)_n \xrightarrow{h\nu} M(CO)_{n-1} + CO$$

$$K + M(CO)_{n-1} \xrightarrow{h\nu} K^{+} + M(CO)_{n-1}^{-}.$$

The anions $Cr(CO)_5^{-}$ and $W(CO)_5^{-}$ have been observed and assigned a C_{4v} structure. It will be noted that these are odd-electron species and should give rise to e.s.r. spectra.

8.6 Photolysis in a reactive matrix

Carbon monoxide is a two-electron ligand, so there are few stable mononuclear carbonyls of odd-electron transition metals. Vanadium forms the readily reduced $V(CO)_6$, but manganese (7e) and cobalt (9e) form only binuclear carbonyls $Mn_2(CO)_{10}$ and $Co_2(CO)_8$. It is possible, however, to form mononuclear complexes of these elements with NO, a three-electron ligand, manganese forming $Mn(CO)_4NO$, isoelectronic with $Fe(CO)_5$, and cobalt forming $Co(CO)_3NO$, isoelectronic with $Ni(CO)_4$.

If these nitrosyl complexes are photolysed *in a carbon monoxide matrix* NO is lost and replaced by CO, so that novel odd-electron carbonyls are formed. $Mn(CO)_4NO$ is said to give $Mn(CO)_4$, $Mn(CO)_5$ and $Mn(CO)_6$ on photolysis in carbon monoxide, while $Co(CO)_3NO$ gives $Co(CO)_4$. The last species is thought to be less symmetric than $Ni(CO)_4$ (which has T_d symmetry) as it appears to give rise to two carbonyl stretching bands rather than one.

The reverse reaction occurs if matrices heavily 'doped' with NO are used; thus photolysis of $Os(CO)_5$ in an argon/NO matrix gives bands assigned to $Os(CO)_2(NO)_2$, analogous to the stable iron compound $Fe(CO)_2(NO)_2$.

Even nitrogen is reactive towards these carbonyl fragments; photolysis of $Ni(CO)_4$ in a nitrogen matrix gives bands attributed to $Ni(CO)_3N_2$, and several similar reactions have been reported.

8.7 Metal atom co-condensation with reactive matrices

This powerful technique, which has been used with great success for the large-scale production of compounds by co-condensation, has also been adapted to allow the investigation of the formation of intermediates, largely by infrared and Raman spectroscopy. When transition metals are used very high temperatures are needed for evaporation of metal atoms, so the precautions noted in chapter 4 to prevent over-heating of the matrix are necessary. The basic technique is otherwise simple; a beam of evaporated metal atoms is co-condensed with an excess of matrix gas, as in a normal co-condensation experiment. However, instead of an inert matrix a reactive matrix, composed wholly or partly of molecules able to react with the metal atoms, is employed.

Carbon monoxide matrices. Some of the earliest experiments involved the use of carbon monoxide or carbon monoxide-doped matrices. Nickel atoms were shown to react during or after deposition, the products depending on the proportion of carbon monoxide in the matrix. A matrix containing one part of carbon monoxide in 500 of argon gave initially a single new band at 1996 cm^{-1} assigned to

NiCO. Annealing, which allowed diffusion of carbon monoxide to occur, led to the gradual appearance of other bands and the corresponding decrease of the first band, suggesting that $Ni(CO)_2$, $Ni(CO)_3$ and eventually $Ni(CO)_4$ were being formed. Matrices containing higher proportions of carbon monoxide gave the higher carbonyls initially, while a pure carbon monoxide matrix, as might be expected, led to formation of $Ni(CO)_4$ only.

$Ni(CO)_4$ is, of course, one of the best known and most readily formed carbonyls, and its formation is thus hardly surprising. The heavier members of the same group, palladium and platinum, on the other hand, are not known to form stable simple carbonyls. It is therefore most interesting that the co-condensation of palladium atoms with carbon monoxide or argon/carbon monoxide matrices leads to the appearance of 'carbonyl bands' exactly analogous to those for the nickel carbonyls (see table 8.7). The mono- and dicarbonyls appear to be linear while the tricarbonyls appear to be planar (D_{3h}), since only one carbonyl stretching frequency was observed in each case. There is general agreement with the independent photolysis work except in the uncertainty as to whether $M(CO)_3$ is planar, as expected from electron-pair repulsion theory, or pyramidal.

$Pt(CO)_4$ is the only platinum carbonyl reported in the matrix, but it, like $Pd(CO)_4$, is not stable enough to survive heating to room temperature despite an '18-electron' configuration.

Turning to earlier transition groups of the periodic table, we may note the formation of mononuclear carbonyls of tantalum ($Ta(CO)_x$ with $x = 1$ to 6) and of carbonyls of chromium, molybdenum and tungsten with $x = 1$ to 4 by the co-condensation method. It seems more than likely that most, if not all of the groups V–VIII transition metals will be found to form carbonyls in these circumstances.

Carbonyls of non-transition elements

More recent work has produced evidence for the existence of carbonyls of elements *outside* the transition metal groups, and of elements of the lanthanide and actinide groups. These results do not, of course, disprove the hypothesis that $d \rightarrow \pi^*$ back-bonding is important in stabilising the 'conventional' transition metal carbonyls; the new products are extremely unstable and are formed under conditions where no alternative modes of compound formation are possible. On the other hand, $d \rightarrow \pi^*$ back-bonding to the antibonding orbital in carbon monoxide explains the reduction in the C–O stretching frequency, so it is perhaps surprising that non-

TABLE 8.7. *Ni, Pd and Pt carbonyls: frequencies/cm^{-1}*

M	MCO	$M(CO)_2$	$M(CO)_3$	$M(CO)_4$
Ni	1996	1967	2017	2052
Pd	2050	2044	2057	2071
Pt				2055

transition metal carbonyls give C–O stretches in exactly the same region as their transition metal counterparts.

The coinage metals, copper, silver and gold, though usually considered to be transition metals, have filled d-shells in the M^0 states, and no simple carbonyls have been prepared in normal circumstances. Co-condensation experiments in pure carbon monoxide matrices, however, show that carbonyl formation can occur; the two IR bands observed for copper and silver probably arise from pyramidal $M(CO)_3$ molecules. These should have an unpaired electron and give rise to e.s.r. spectra.

Uranium, with a partly filled 5f-subshell, and two lanthanides, neodymium and ytterbium, with partly filled 4f-subshells, also give rise to carbonyl stretching bands in the IR spectrum when the metals are co-condensed with carbon monoxide-doped argon matrices. Species up to $U(CO)_6$ and $Nd(CO)_6$ have been suggested. Again, one may expect further developments in this field.

The most striking results, perhaps, are those which suggest that the pre-transition element, aluminium, can form carbonyls. Aluminium can have no occupied d-orbitals, yet co-condensation of aluminium with 3 % carbon monoxide in krypton gives two carbonyl bands at 1988 cm^{-1} and 1890 cm^{-1}, assigned to bent $Al(CO)_2$. This assignment is supported by mixed $C^{16}O/C^{18}O$ results.

The post-transition elements germanium and tin give bands assigned to MCO and higher carbonyls. The 'carbonyl stretch' in GeCO is said to be at 1908 cm^{-1}; the molecule, like CCO, should have a $^3\Sigma$ ground state with two unpaired electrons. It seems unlikely that back-donation from the filled 3d-orbitals occurs, as these orbitals are fairly deep (binding energy ~ 28 eV); it is more than likely though that overlap of the π^* orbital of the carbonyl with the partially occupied 4p-level of germanium occurs, weakening the C–O bond and giving a lowered carbonyl frequency.

Nitrogen matrices. Nitrogen is isoelectronic with carbon monoxide, but only a few 'dinitrogen complexes' have been prepared by conventional methods. Nevertheless, co-condensation of metal atoms with nitrogen matrices has shown that nitrogen can form a wide range of compounds with transition metals, many of which are similar to carbonyl complexes.

The most extensive work to date has involved nickel atoms co-condensed with pure nitrogen or argon/nitrogen matrices. A total of four species, ranging from NiN_2 to $Ni(N_2)_4$, are observed in IR and Raman studies. One of the main interests in 'dinitrogen' complexes was whether the nitrogen was bonded 'end-on' like carbonyl or in a sideways fashion. If mixed $^{14}N/^{15}N$ studies are carried out then 'end-on' NiN_2 should give rise to four different isotopic species, whereas 'sideways' NiN_2 should give only three since the nitrogen atoms are equivalent. It has been found that the nickel dinitrogen species all involve 'end-on' bonding and their structures (see fig. 8.3(*a*)) are similar to the corresponding carbonyls.

Palladium and platinum form $M(N_2)_n$ (n = 1–3) which again appear to correspond to their carbonyls. Copper also forms a dinitrogen species, but it is not clear whether a single metal atom or a cluster is involved. In cobalt/nitrogen studies, isotopic mixtures ($^{14}N_2$, $^{14}N^{15}N$, $^{15}N_2$) give a triplet rather than a quartet, suggesting that sideways bonding may occur (see fig. 8.3(*b*)).

Oxygen complexes. Nickel, palladium and platinum all form two species MO_2 and $M(O_2)_2$ when their atoms are co-condensed with matrices containing oxygen. $^{16}O_2$, $^{16}O^{18}O$, $^{18}O_2$ studies show that in MO_2, the oxygen atoms are equivalent. Thus oxygen, unlike nitrogen, coordinates sideways to nickel. Similar bonding occurs in the $M(O_2)_2$ species, and a unique D_{2d} structure is proposed (see fig. 8.3(*c*)). The oxygen atoms form a tetrahedron around the nickel, suggesting that no more than two oxygen molecules can be coordinated.

Mixed complexes. A number of transition metals have been found to give mixed carbonyl/nitrogen complexes when the atoms are co-condensed with matrices containing carbon monoxide and nitrogen. The nickel system is again the best characterised and all $Ni(N_2)_x(CO)_{4-x}$ (x = 0–4) are observed. It is probably fairly general that nitrogen (if present) can replace at least one carbonyl, during transition atom co-condensations, and form a mixed complex. These compounds may explain many of the unassigned bands reported in

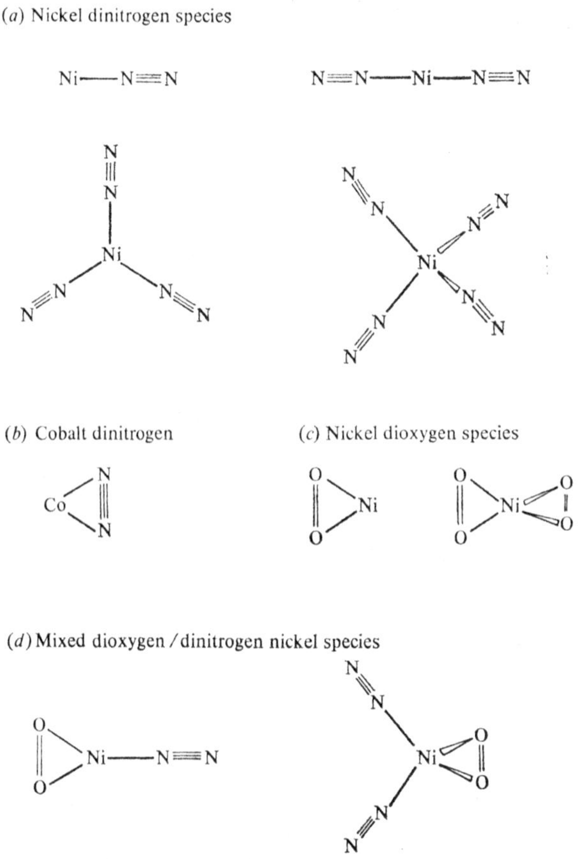

Fig. 8.3. Dinitrogen and dioxygen species of transition metals.

early studies involving evaporation of metals, as slight nitrogen leaks are common and usually disregarded since nitrogen is not seen in the IR, and carbon monoxide is often released from inadequately degassed high temperature surfaces.

Mixed oxygen/nitrogen complexes are also found for nickel; NiN_2O_2 and $Ni(N_2)_2O_2$ (see fig. 8.3(*d*)) are the only two expected and they have both been observed.

Note added in proof (see p. 121). The band at 478 cm^{-1} for MgF_2 systems has now been shown to be due to some other species, probably a dimer; the whole analysis of this and the other systems must therefore be regarded as unsound, and the bond angles reported in table 8.3 should quite possibly be taken as 180°, as expected.

9 Conclusions and outlook

We have outlined the development of the technique of matrix isolation and the special aspects of cryogenics, vacuum technology and spectroscopy that are involved in the production and investigation of matrix-isolated species. The main technical difficulties have already been overcome, and matrix isolation is within the reach of most university chemistry departments. We have also given an account of some of the results that have been published so far, results that have established the usefulness of matrix isolation in some major areas of chemistry. It remains for us to indicate what we see as the most likely technical and spectroscopic developments in the near future, and the areas of chemistry in which matrix isolation may have most to contribute.

Technically, the modern microrefrigerator is probably as convenient for most matrix studies as possible, and little advance in this direction seems likely. Similarly, there is unlikely to be any significant development in the vacuum technology associated with matrix isolation. The main technical developments are likely to be those associated with the spectroscopic aspects – improvements of sensitivity so that weak signals from very dilute samples can be detected, and the use of techniques other than infrared, UV/visible and e.s.r. spectroscopy.

The first breakthrough in this direction has already occurred, in that modern techniques of Raman spectroscopy make it possible for the Raman spectra of matrix-isolated species to be obtained; further advances in sensitivity are likely in this field.

Similarly, the most modern far-infrared spectrometers, which work on the interferometer principle, make it possible now to record the strongest bands of matrix-isolated species in the far-infrared region (below $\sim 250\ cm^{-1}$). Again, it seems likely that rapid advances will be reported here.

The 'newer' spectroscopic methods, such as photoelectron spectroscopy, magnetic circular dichroism and Mössbauer spectroscopy will have some results to report, but for various reasons, some of which we have mentioned in chapter 5, it is unlikely that they will contribute greatly to the investigation of matrix-isolated species.

Of the methods used to prepare matrix-isolated species, those involving reactive matrices offer the greatest hopes for future development. The range of possible chemistry here is very wide, particularly if the spectroscopic problems associated with the use of larger molecules as matrix materials can be overcome.

The use of furnaces for evaporation and reaction of volatile fragments from solids will become increasingly important now that the usefulness of matrix isolation in characterising the products of such processes is established. Other methods of producing transient species that may be subsequently trapped in matrices, such as discharge reactions, seem by contrast to be capable of less further development, though they will continue to prove useful as in the past.

Methods involving photolysis of matrix isolated precursors will continue to be important, and further advances in the production of ions in matrices by photo-ionisation may be expected. A full study of the modes of fragmentation of a precursor as the energy of the exciting radiation is changed could be very important as an exercise in photochemistry, and we may expect matrix isolation to contribute more to such studies in the future. In particular, studies of photolysis and photo-ionisation of organic molecules in this way could be extremely informative.

It should be possible, even now, for almost any small molecule to be prepared 'to order' in a matrix by some one or more of the techniques we have discussed in chapter 4, unless it is too reactive to survive even under matrix conditions. This freedom to prepare 'tailor-made' reactive or unstable molecules for study at leisure is obviously one of the most important advantages of matrix isolation. This has been demonstrated in the production of a whole range of simple metallic carbonyl, nitrogen and oxygen species which has opened a new field of chemistry and given considerable insight into the bonding and reactivity of metal atoms.

There remain two vitally important aspects of matrix isolation studies that must be investigated further. The problems associated with characterising matrix-isolated species from their spectra, and the effects of the matrix on a species and its spectrum, must be resolved. Secondly, the *use* of matrix isolation to study chemical reactions by characterisation of their products must be developed.

The study of stable molecules, and of other molecules whose spectra are well characterised in the gas phase, is of course most important in connection with the first of these. Further studies involving characterisation of all the fundamentals and some overtones and combinations (in the vibrational spectrum) for stable molecules will help to establish clearly the effects of matrices on molecular potential energy surfaces. This should make it possible for sensible assignments to be made of the spectra of unstable species whose gas-phase spectra are unknown.

It should be noted in passing that the ready availability of matrix isolation equipment now makes it reasonable to use the technique for the spectroscopic characterisation of *stable* molecules. The infrared spectrum in particular is so much simplified by the loss of rotational fine-structure that weak and close bands can be readily detected, and the precise measurement possible on peaks in the spectra of matrix-isolated molecules allows analysis of mixtures to be carried out on such samples.

The subjects of matrix shifts and matrix splittings seem to need further study, though much is already known about their nature. Again, studies of stable molecules are most likely to help in these respects.

The final area, that of application to chemical reactions, in which developments must be expected is, of course, the most important. Like all techniques, matrix isolation is a tool that must be put to use to remain vital. The work that has been done so far has been partly concerned with the characterisation of the tool, and partly concerned with its application to some interesting areas of chemistry. We believe that it is in the broadening of the field of application of matrix isolation that the most important future advances will lie.

Bibliography

Reviews

A. J. Downs and S. C. Peake, *Molecular Spectroscopy*, vol. 1, Specialist Periodical Reports of the Chemical Society, London, 1973, pp. 523–58.

J. S. Ogden and J. J. Turner, *Chem. in Britain*, 1971, **7**, 186.

W. Weltner Jr, *Adv. High Temp. Chem.*, 1969, **2**, 85.

G. C. Pimentel and S. W. Charles, *Pure Appl. Chem.*, 1963, **7**, 111.

Books

Formation and Trapping of Free Radicals, ed. A. M. Bass and H. P. Broida, Academic Press, New York, 1960.

B. Meyer, *Low Temperature Spectroscopy*, Elsevier, New York, 1971.

Some important papers

E. Whittle, D. A. Dows and G. C. Pimentel, *J. Chem. Phys.*, 1954, **22**, 1943. (Proposal of matrix isolation method.)

M. J. Linevsky, *J. Chem. Phys.*, 1961, **34**, 587. (Studies of evaporated solid (LiF).)

R. L. Morehouse, J. J. Christiansen and W. Gordy, *J. Chem. Phys.*, 1966, **45**, 1751. (E.s.r. of radicals MH_3.)

D. E. Milligan and M. E. Jacox, *J. Chem. Phys.*, 1967, **47**, 5146. (IR and UV of photolysis products (CH_3 from CH_4).)

L. Andrews, *J. Chem. Phys.*, 1969, **50**, 4288. (Use of reactive matrix (Li in oxygen/argon).)

R. L. DeKock, *Inorg. Chem.*, 1971, **10**, 1205. (Use of reactive matrix (Ni and Ta in CO/argon).)

Index